Practical Switching
Power Supply Design

Marty Brown
Motorola Semiconductor

MOTOROLA
Series in Solid State Electronics

Practical Switching Power Supply Design

ACADEMIC PRESS, INC.
Harcourt Brace Jovanovich, Publishers
San Diego New York Boston London Sydney Tokyo Toronto

This book is printed on acid-free paper. ∞

ACADEMIC PRESS, INC.
1250 Sixth Avenue
San Diego, California 92101

United Kingdom Edition published by
Academic Press Limited
24–28 Oval Road, London NW1 7DX

Library of Congress Cataloging-in-Publication Data

Brown, Marty.
    Practical switching power supply design / Marty Brown.
       p. cm.
    ISBN 0-12-137030-5 (alk. paper)
    1. Switching circuits--Design and construction. 2. Power
semiconductors--Design and construction. 3. Semiconductor switches-
  -Design and construction.   I. Title.
    TK7868.S9B66   1990
    621.381'5--dc20   89-17518
                                           CIP

PRINTED IN THE UNITED STATES OF AMERICA
92 93 94 95 96 QW 9 8 7 6 5 4 3

Contents

CHAPTER 5

Semiconductors Used in a Switching Power Supply   43

CHAPTER 6

The Magnetic Components within a Switching Power Supply   67

CHAPTER 7

Cross-Regulation of the Outputs   97

CHAPTER 8

Protection   103

CHAPTER 12

Switching Power Supply Design Examples   199

Preface

In the age of specialization for electronics engineers, it becomes very difficult to maintain a level of competence within a broad range of electronics fields. Nonetheless, many engineers will be assigned design projects outside their primary field of expertise, among which are switching power supplies. This is done primarily because the engineer has a unique ability to learn technical subjects relatively quickly. Unfortunately, the literature available today on the subject of switching power supplies tries to convey an understanding through lengthy derivations of applied mathematics. This does not work since only an intuitive sense of the subject matter creates an understanding of the fundamental relationships.

This book is written for just this purpose. It contains *no* mathematical derivations. Instead, it contains written explanations in semitechnical terms, on such topics as magnetic behavior and feedback compensation, to give the reader a good intuitive understanding of the operation of a switching regulator. The material highlights the areas that have a strong bearing on the supply's reliable operation that are not obvious from the "paper design." It also attempts to tie together the often oblique and unrelated information presented in component manufacturer's catalogs. The design examples are written in a clear step-by-step fashion in order to show the reader the steps necessary in a typical switching regulator design. They were also chosen because of their utility in a wide range of typical applications. They can be easily modified and scaled to fit many more applications. The topics contained in the book range from considerations in capacitor and semiconductor selection to quasi-resonant converter design.

This book has been written as a result of many years of learning about switching power supplies from experience and equally many years of frustration with the available technical resources. The material is organized specifically to answer those questions that I and the many engineers with whom I have conversed have had when faced with a switching power supply design. In short, this material is written *for* a working engineer *by* a working engineer.

1

Why Use Switching Power Supplies?

The choice of whether to use a switching or linear power supply in a particular design is significantly based on the needs of the application itself. Both linear and switching supplies have distinct merits in certain areas of application. So in order to use the more appropriate power supply type in a particular design, it is necessary to understand the cost and electrical requirements of the entire product and select the type of power supply that best satisfies those requirements.

The linear power supply offers the designer three major advantages. The first advantage is its simplicity. One can purchase an entire linear regulator in a package and simply add two filter capacitors for storage and stability. Even if the designer had to design a linear regulator from scratch, the present technology is sufficiently mature to allow duplication of a design from a book with very little effort. The second major advantage is its quiet operation and load-handling capability. The linear regulator generates little or no electrical noise on its output, and its dynamic load response time—the time it takes to respond to changes in the load current—is very short. The third advantage is that, for an output power of less than approximately 10 W, its component costs and manufacturing costs are less than the comparable switching regulator.

The disadvantages of the linear-type regulator are what limit its range of application. The designer cannot eliminate these shortcomings but can attempt to minimize their effects. First, it can be used only as a step-down regulator, which means that the designer must somehow develop an input voltage that is at least 2 to 3 V higher than the required output voltage. This means that in off-line situations a 50–60-Hz transformer with rectification and filtering must be placed before the linear power supply. This pre-power conditioning increases the system cost. Second, each linear regulator can have only one output. So for each additional output voltage required, an entire separate linear regulator must be added. This requirement for multiple voltages once again drives up the system cost. Another major disadvantage is the average efficiency of linear regulators. In normal applications, linear regulators exhibit efficiencies of 30 to 60 percent. This means that for every watt delivered to the load, more than one watt is lost within the supply. This loss, called the *headroom loss*, occurs in the pass transistor and is, unfortunately, necessary to develop the needed biases within the supply required for operation and varies greatly when the input voltage varies between its high- and low-line specifications. This makes it necessary to add heat-sinking to the pass transistor that will be sufficient to handle the lost power at the highest specified input voltage and the highest specified load current. Most of the time the supply will not be operating under these circumstances, which means that the heatsink will be oversized during most of its operating life. This once again is an added system cost. The point where the heatsink cost begins to become prohibitive is about 10 W of output power. Up to this point, any convenient metal structural member can adequately dissipate the heat. These shortcomings greatly escalate at higher output power levels and quickly make the switching regulator a better choice.

The switching regulator circumvents all of the linear regulator's shortcomings. First, the switching supply exhibits efficiencies of 68 to 90 percent regardless of the input voltage, thus drastically reducing the size requirement of the heatsink and hence its cost. The power transistors within the switching supply operate at their most efficient points of operation: saturation and cutoff. This means that the power transistors can deliver many times their power rating to the load and the less expensive, lower-power packages can be used. Since the input voltage is chopped into an AC waveform and placed into a magnetic element, additional windings can be added to provide for more than one output voltage. The incremental additional cost of each added output is very small compared to the entire supply cost—and in the case of transformer-isolated switching supplies, the output voltages are independent of the input voltage. This means that the input voltage can vary above and/or below the level of the output voltages without affecting the operation of the supply. The last major advantages are its size and cost at the higher output power levels. Since their frequency of operation is very much greater than the 50–60 Hz line frequency, the magnetic and capacitive elements used for energy storage are much smaller and the cost to build the switching supply becomes less than the linear supply at the higher power levels.

All of these advantages make the switching power supply a much more versatile choice, with a wider range of applications, than the linear supply.

The disadvantages of the switching supply are minor and usually can be overcome by the designer. First, the switching supply is more complicated than the comparable linear supply. If a switching supply cannot be bought off-the-shelf to suit the needs of the product, then it must be designed. At this point the time it takes to design a reliable switching supply to suit one's needs can be quite sizable, and if this is the first power supply design undertaken by the designer, the learning curve can add significantly to this time. Don't be lulled into believing that the design is "cookbook." Many more considerations must be taken into account even if there is a published design that will meet the needs of the product. The experienced power supply designer will need a minimum of three worker-months, depending on its complexity, to design, prototype, and test the supply before releasing it to production. It is safe to plan on 4 to 6 worker-months' worth of effort to perfect the design prior to production. Obviously this design effort comes at a cost, and this must be considered during the product planning stage of the program. Second, considerable noise from the switching supply is generated on its outputs and input and radiated into the environment. This can be difficult to control and certainly cannot be ignored during the design phase. A little knowledge of radio-frequency (RF) behavior and design can go a long way in aiding the engineer during the design phase. There can be simple solutions to this problem, but generally additional filtering and shielding will have to be added to the supply to limit the effects of the noise on the load and the environment. This, of course, adds cost to the supply. Third, since the switching supply chops the input voltage into time-limited pulses of energy, the time it takes the supply to respond to changes in the load and the input is slower than the linear power supply. This is called *transient response time*. To compensate for this sluggishness, the output filter capacitors usually must be increased in value to store the energy needed by the load during the time the switching supply is adjusting its power throughput. Once again added cost is incurred, but note that all of these disadvantages are under the control of the designer and their impact on the supply and the system can be minimized.

Generally, the industry has settled into areas where linear and switching power supplies are applied. Linear supplies are chosen for low-power, board-level regulation where the power distribution system within the product is highly variable and the load's supply voltage needs are restricted. They are also used in circuits where a quiet supply voltage is necessary, such as analog, audio, or interface circuits. They are also used where a low overhead cost is required and heat generation is not a problem. Switching power supplies are used in situations where a high supply efficiency is necessary and the dissipation of heat presents a problem, such as battery-powered and handheld applications where battery life and internal and external temperatures are important. Off-line supplies are also typically switchers because of their efficiency in generating all the voltages needed within the product, especially in very-high-power applications, up to many kilowatts.

In summary, because of its versatility, efficiency, size, and cost, the switching power supply is preferred in most applications. The advances in component technology and novel topological design approaches will only add to the desirability of the switching power supply in most applications.

2

How a Switching Power Supply Works

Conceptually, switching regulators are not difficult to understand. When viewed as a blackbox with input and output terminals, the behavior of a switching regulator is identical to that of a linear regulator. The fundamental difference is that a linear regulator regulates a continuous flow of current from the input to the load in order to maintain a constant load voltage. The switching regulator regulates this same current flow by chopping up the input voltage and controlling the average current by means of the duty cycle. When a higher load current is required by the load, the percentage of on-time is increased to accommodate the change.

Two basic types of switching regulators constitute the foundation of all of the pulsewidth-modulated (PWM) switching regulators. These types are the forward-mode regulators and the flyback-mode regulators. The name of each type is derived from the way the magnetic elements are used within the regulator. Although they may resemble each other schematically, they operate in quite different fashions.

2.1 Forward-Mode Switching Regulators

Forward-mode switching regulators have as their functional components four elements: a power switch for creating the PWM waveform, a rectifier (or catch diode), a series inductor, and a capacitor (see Fig. 2.1). The power switch may be a power transistor or a metal oxide semiconductor field-effect transistor (MOSFET) placed directly between the input voltage and the filter section. In between the power switch and the filter section there may be a transformer for stepping up or down the input voltage as in transformer-isolated forward regulators. The shunt diode, series inductor, and shunt capacitor form an energy storage res-

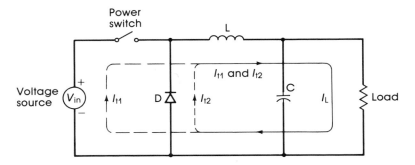

Figure 2.1
Forward regulator and its current flow.

ervoir whose purpose is to store enough energy to maintain the load voltage and current over the entire off-time of the power switch. The power switch serves only to replenish the energy lost to the load during its off-time. Its function can be seen as an electrical equivalent of a mechanical piston–flywheel combination. The piston provides a pulse of energy, and the flywheel stores the mechanical energy for use by the load.

The operation of the power switch can be broken up into two periods. The first is when the power switch is on. During this period, the load current passes from the input source, through the inductor to the load, and back again through the return (or ground) lines to the input source. During this time the diode is reverse-biased. After the power switch turns off, the inductor still expects current to flow through it. The former current path through the input source is now open-circuited, and the catch diode now begins to conduct, thus maintaining a closed current loop through the load. When the power switch once again turns on, the voltage presented to the filter serves to turn off the catch diode. In short, forward current is always flowing through the inductor; hence its name.

The amount of energy being delivered to the load is controlled by the duty cycle of the power switch on-time. This may vary anywhere between 0 and 100 percent duty cycle but typically falls between 5 and 95 percent duty cycle. An approximate model of the relationship between input voltage, duty cycle, and output voltage is that the output voltage is the average of the area under the chopped voltage waveform or

$$V_{\text{out}} \cong V_{\text{in}} \cdot \text{duty cycle} \tag{2.1}$$

In reality this relationship applies only for light loads, but it does serve as a reasonable approximation elsewhere.

2.2 Flyback-Mode Switching Regulators

Flyback-mode switching regulators have the same four basic elements as the forward-mode regulators except that they have been subtly re-arranged (see Fig. 2.2). Now the inductor is placed directly between the input source and the power switch. The anode lead of the rectifier is placed on the node where the power switch and inductor are connected, and the capacitor is placed between the rectifier output (cathode) and ground (return).

The flyback's operation can be broken up into two periods. When the power switch is on, current is being drawn through the inductor, which causes energy to be stored within its core material. The power switch then turns off. Since the current through an inductor cannot change instantaneously, the inductor voltage reverses (or flies back). This causes the rectifier to turn on, thus dumping the inductor's energy into the capacitor. This continues until all the energy stored in the inductor during the previous half-cycle is emptied. Since the inductor voltage flies back above the input voltage, the voltage that appears on the output capacitor is higher than the input voltage. Note that the only storage for the load is the output filter capacitor. This makes the output ripple voltage of flyback converters worse than their forward-mode counterparts.

The duty cycle in an elementary flyback-mode supply is 0 to 50 percent. This restriction is due to the time required to empty the inductor's flux into the output capacitor. Duty cycles within transformer-isolated flyback regulators can sometimes be larger because of the effects of the turns ratio and the inductances of the primary and the secondary.

The relationship of the output voltage to the input voltage is slightly more difficult to describe. During the power switch's off-time, the in-

Figure 2.2

Flyback-mode regulator and its current flow.

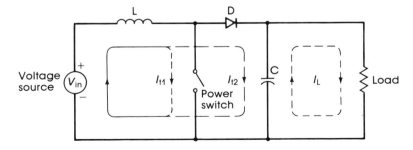

ductor will empty itself before the start of the next power switch conduction cycle. Since the volt-time products of the inductor charging and
discharging cycles must be equal and the output for a nonisolated
"boost" converter must be higher than the input voltage, the resulting
relationship is

$$V_{in} \cdot T_{on} = V_{fLBK} \cdot T_{FLBK}$$

$$V_{out} \cong V_{in} + V_{flbk} \cong V_{in}\left(1 + \frac{T_{on}}{T_{flbk}}\right) \tag{2.2}$$

At the minimum operating voltage, the duty cycle reaches 50 percent
and T_{flbk} equals the total operating period minus the "on-time."

3

A Walk through a Representative Switching Power Supply

In order to adequately approach a switching power supply design, the designer must have a reasonable understanding of the major subsections that make up a switching power supply. The subsections discussed represent a typical minimum system. Additional functionality may be added to the supply by adding to these basic subsections. The supply discussed is a single output, push–pull regulator. The circuit sections and waveforms are shown in Figures 3.1 and 3.2.

3.1 The EMI Filter

This section is composed of a small L–C filter between the input line and the regulator. It serves a dual purpose. First, C_1 and L_1 act as a high frequency radio-frequency interference (RFI) filter, which reduces the conducted high frequency noise components leaving the switching supply back into the input line. These noise currents would then radiate from the input power lines as in an antenna. The lowpass cutoff frequency of this filter should be no higher than 2 to 3 times the supply's operating frequency. The second purpose of this stage is to add a small impedance (L_1) between the input line and the bulk input capacitor. It basically reduces any lethal transient voltage and allows the bulk input filter capacitor and any surge protector to absorb the destructive energies from the input line spikes or surges with little chance of exceeding any of the components' voltage ratings.

3.2 Bulk Input Filter (Storage) Capacitor

This capacitor is relatively large in value. It has the responsibility of storing the high- and low-frequency energy required by the supply dur-

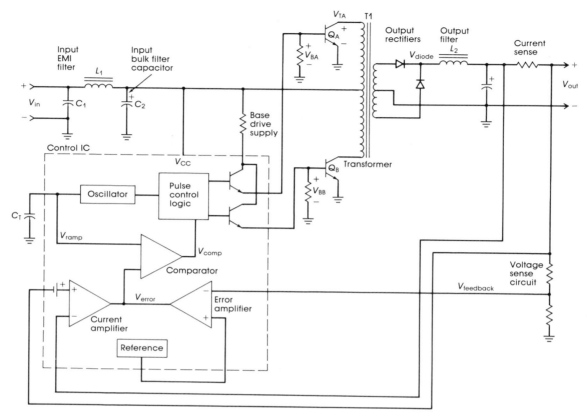

Figure 3.1

A walk through a representative switching regulator circuit.

ing each power transistor's conduction cycle. It is usually made up of at least two capacitors, an electrolytic or tantalum capacitor for the current components at the supply's switching frequency and a ceramic capacitor for the switching frequency harmonics. This capacitance must represent a low impedance from direct current (DC) to many times the switching frequency of the supply. Another factor that necessitates the use of the bulk input capacitor is that the input line may have long lengths of wire or printed circuit board trace, which adds series resistance and inductance between the power source and the supply. The input line at high frequencies actually resembles a current-limited current source and cannot deliver the high-frequency current demands of the supply necessary for the fast voltage and current transitions within the supply. The input capacitor charges at a low frequency and sources current over a much higher frequency range. Without both a low-frequency electrolyte-

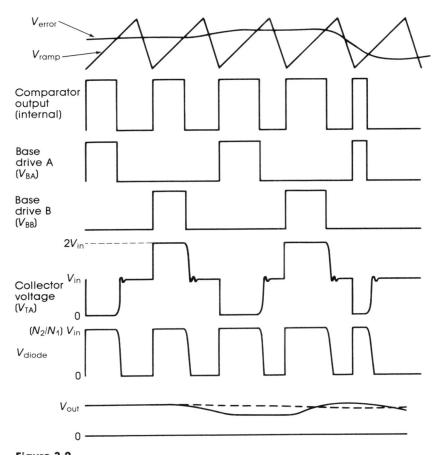

Figure 3.2
Representative waveforms.

type capacitor and a high-frequency ceramic-type capacitor, the supply would starve for high-frequency current and voltage and adversely affect the supply's stability.

3.3 Transformer

In this configuration, the transformer provides DC isolation between the input line and the output. It also performs a voltage step-up and/or step-down function for the supply. The transformer does not store energy in this configuration. Additional outputs may be added by simply adding another winding on the secondary. This allows one switching supply to provide all the voltages required by most product designs. The transformer is also the backbone of the switching power supply. If the transformer is improperly designed, it would adversely affect the supply operation and the reliability of the semiconductors.

3.4 Power Switches

These are power transistors or MOSFETs that are quickly switched between the saturation (full-on) and the cutoff (full-off) states. They serve as a "gate" for the energy entering the supply that is subsequently delivered to the load. The energy flow is regulated by the control circuit, which senses the energy demanded by the load and then varies the percentage of on-time for the power switches, which then "regulate" the delivery of the energy to the load to match the load's demands. The power switches also represent the least reliable components within the supply. If any components are to fail during an adverse operating condition, these would be the first ones to fail. So great care should be taken during the design and selection phase to ensure their reliable performance.

3.5 Output Rectifiers

In this regulator configuration, the output rectifiers conduct at the same time as the power switches. The secondary voltage waveforms in isolated configurations such as this have an average DC value of zero (centered about 0 V), but during the on-time of the power switches the secondary voltage reaches peak values of the turns ratio times the input voltage. The rectifiers convert this bipolar waveform into a unipolar pulse train. To change the polarity of the output voltage, one simply reverses the rectifier's polarity. Although the rectifier conducts an average current equal to the load current, the peak value of the current will be higher than the average. So during the rectifier selection process the designer should consider any additional losses incurred during these high peak currents and add a margin to the current specification.

3.6 The Output Filter Section

This is an example of the output filter section of a forward-mode converter. This filter is called a *choke input filter* (or *LC* filter) and is a series inductor followed by a shunt capacitor. Its purpose is to store energy for the load during the times when the power switches are not conducting. It basically operates like an electrical equivalent of a mechanical flywheel. The on-time of the power switches serves only to replenish the energy lost by the inductor during their off-time. Typically, approximately 50 percent more energy is stored in the inductor and capacitor than is needed by the load over the entire period. This reserve can be drawn on by sudden increases in load demand until the control loop can provide more energy by increasing the on-time of the power switches.

3.7 Current Sense Elements

The method shown here is only one way of implementing the overcurrent sensing function. Essentially, the purpose is to develop a voltage that is proportional to the output load current. This voltage is then amplified, and if it becomes too high (an overcurrent condition), it overrides the voltage regulator control loop and forces a reduction in the output voltage. Depending on the way the output current is sensed, what other parameters are summed in, and the gain of the current-sensing amplifier, one can either achieve a constant power limiting, a constant current limiting, or a current foldback limiting. The type of limiting chosen depends on how much power the load can withstand during an overcurrent or short-circuit failure. In voltage-mode regulators this feature remains completely inactive until an abnormal overcurrent condition is entered. In current-mode control regulators, the transformer's primary current is sensed and used as part of the overall control strategy of the supply, offering not only overcurrent protection but also improved supply responsiveness.

3.8 Voltage Feedback Elements

This is usually a resistor divider, which reduces the rated output voltage to the same voltage appearing as the reference voltage on the input to the voltage error amplifier. The voltage error amplifier amplifies the difference between the ideal level—dictated by the reference voltage—and the actual output voltage as presented by the feedback elements and controls the on-time of the power switches accordingly.

3.9 The Control Section

This function is typically centered around a switching power supply control integrated circuit. It performs the functions of DC output voltage sensing and correction, voltage-to-pulsewidth conversion, a stable reference voltage, an oscillator, overcurrent detection and override, and the power switch driver(s). It may also include a soft-start circuit, dead-time limiting, and a remote shutdown. The oscillator sets the frequency of operation of the supply and generates a sawtooth waveform for the DC-to-pulsewidth converter. The voltage error amplifier amplifies the difference between the "ideal" reference voltage and the sensed output voltage presented by the resistor divider feedback elements. The error amplifier's output voltage represents this error between the reference and the actual output multiplied by the high DC gain of the operational amplifier riding on a DC offset. This error signal is then presented to the DC-to-pulsewidth converter, which produces a pulsetrain whose duty cycle represents this error signal. This pulsetrain is then presented to the power switch driver(s). If the supply is single-ended, that is, has only one power switch, the waveform is used to drive the output driver directly. If it is a double-ended supply (two power switches), this pulsetrain is first placed into a digital flip-flop that steers the pulses alternately between two output drivers. The output drivers themselves usually take one of two forms. First is the uncommitted transistor, which is where both the emitter and collector of the output transistor are brought out of the integrated circuit (IC) and are better suited for driving bipolar power transistor power switches. The second type is the push–pull driver. This type is preferred for driving power MOSFETs. These control functions represent the minimum functionality of a control IC.

Added functionality, which varies from IC to IC, should be considered carefully, keeping in mind the system design. This might include soft-start, remote shutdown, and synchronization. Soft-start reduces the inrush current into the supply during startup by overriding the error amplifier and hard-limiting the initial maximum pulsewidths until the supply has reached its desired output. Remote shutdown is a circuit that inhibits supply operation electrically by shutting down the control functions without removing power to the power sections of the supply. This feature is intended for those applications where it is impractical to interrupt the supply's high-current input line. Synchronization is needed for those systems having sections where the fixed frequency output ripple of the power supply would interfere with a critical system circuit such as a

cathode-ray-tube display or an analog-to-digital or ditigal-to-analog converter. In those cases the conduction pulses would be sychronized in
phase and frequency to the critical circuit and could be placed in phase
such that the critical circuit would be immune to the supply's ripple
voltage. It also may be necessary to synchronize more than one switching power supply. The designer must study each control IC carefully in
order to select the most appropriate IC for the application.

These basic functional subsections represent the minimum functionality that a typical switching power supply should possess. Additional
functions that may or should be added are input transient protection,
undervoltage lockout, output overvoltage protection, and any power se-
quencing that the supply may need to provide to the system. Many items
need to be considered at the system design specification stage of a system development program and should be discussed as early in the program as possible. This will aid the designer in outlining the best possible
design approach to the switching power supply and avoid any last minute design changes downstream in the program.

4

Switching Power Supply Topologies

Switching power supplies gained popularity in the early 1970s, coinciding with the introduction of the bipolar power transistor. The basic theory of the switching power supply has been known since the 1930s. Since the 1930s, many evolutionary changes have occurred to make the switching power supply meet the needs of many diverse applications. For this reason, many variations have evolved, each with merits that make it better suited for particular applications. Some topologies work better at high input voltages, some at higher output power levels, and some are targeted for the lowest cost. Keep in mind that many topologies can work for each particular application, but one topology usually has the right combination of features that makes it the best choice.

4.1 Factors Affecting the Choice of an Appropriate Topology

In order to select an appropriate topology for your application it is necessary to understand the subtle differences between the topologies and what factors make them more desirable for certain applications. Five primary factors differentiate the various topologies from one another:

1. *The peak primary current.* This is an indication of how much stress the power semiconductors must withstand and tends to limit a particular configuration in the output power it can deliver and the input voltage over which it can operate.
2. *How much of the input voltage can be placed across the primary winding of the transformer.* This indicates how effectively power can be derived from the input line. Switching power supplies are constant-power circuits, so the more voltage supplied to the trans-

former or inductor, the less the average and peak currents needed in order to develop the output power.

3. *How much of the B–H characteristic can be used within the transformer during each cycle.* This indicates which configurations have physically smaller transformers for a rated output power.

4. *DC isolation of the input from the load.* This provides DC isolation of the output from the input and allows the designer to add multiple outputs with ease. Transformer isolation may also be necessary in order to meet the safety requirements dictated by the marketplace.

5. *Cost and reliability.* The designer wishes to select a configuration that requires the minimum parts without subjecting the components to undue overstress.

At the beginning of each power supply design effort the designer should perform a little predesign estimation exercise. This is done by making a reasonable assumption about the supply efficiency and working with the general equations involving the peak currents and voltages. From this exercise, one can select the best switching power supply topology, select the preliminary choices for the semiconductors, and even estimate the amount of losses within the components. It may also guide the designer in an approach to packaging the power supply and provide some idea as to the final cost of the supply. This effort can act as an early roadmap during the design phase and also saves time because the designer can order the semiconductor components before the power supply is even designed.

The industry has settled into several primary topologies for a majority of the applications. Figure 4.1 diagrams the approximate range of usage for these topologies. The boundaries to these areas are determined primarily by the amount of stress the power switches (power transistors or MOSFETs) must endure and still provide reliable performance. The boundaries delineated in Figure 4.1 represent approximately 20 A of peak current. Higher peak currents can be used but the power switches would begin to exhibit unusual failure modes, and items such as board layout and lead lengths would become even more critical. It is also no coincidence that these topologies are transformer-isolated topologies. The non-transformer-isolated topologies have very predictable and catastrophic failure modes that most experienced switching power supply designers prefer not to risk.

The flyback configuration is used predominantly for low to medium output power (<150 W) applications because of its simplicity and low

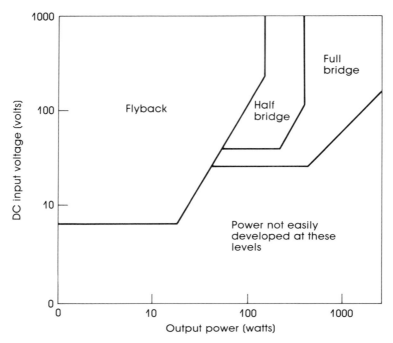

Figure 4.1

Industry favorite configurations and their areas of usage.

cost. Unfortunately, the flyback topology exhibits much higher peak currents than do the forward-mode supplies, so at the higher output powers, it quickly becomes an unsuitable choice. For medium-power applications (100 to 400 W) the half-bridge topology becomes the predominant choice. The half-bridge is more complicated than the flyback and therefore costs more, but its peak currents are about one-third to one-half those exhibited by the flyback. Above 400 W, the peak currents once again become very high and it becomes unsuitable. This is because the half-bridge does not effectively utilize the full power capacity of the input source. Above 400 W the dominant topology is the full-bridge topology, which offers the most effective utilization of the full capacity of the input power source. It also is the most expensive to build, but for those power levels the additional cost becomes a trivial matter. Another topology that is sometimes used above 150 W is the push–pull topology, which exhibits some fundamental shortcomings that make it tricky to use.

By using Figure 4.1 and estimating the major power supply parameters as a preliminary guide at the beginning of a switching power supply design effort, one can be reasonably sure that the final choice of topology will provide a reliable and cost-effective design.

4.2 Non-Transformer-Isolated Switching Power Supply Topologies

The non-transformer-isolated type of switching power supplies are typically used when some external component provides the DC isolation or protection in place of the switching supply. These external components are usually 50–60-Hz transformers or isolated bulk power supplies. Their typical area of application is in local board-level voltage regulation. The non-transformer-isolated supplies are also easy to understand and thus are used as design examples by various manufacturers and subsequently overused by novice power supply designers. Nonisolated-type configurations seldom are used by seasoned power supply designers simply because of the severity of the failure modes caused by the lack of the DC isolation. Also, isolated supplies add a degree of safety by having a second DC dielectric barrier to back up the 50–60-Hz transformer, which enhances the supply's degree of graceful degradation during any possible failures.

There are three basic non-transformer-isolated topologies: the buck (step-down), the boost (step-up), and the buck–boost (inverting). Each topology generates and regulates an output voltage that is above or below the input voltage. Each also has only one output since it is not very practical to add additional outputs to them. Non-transformer-isolated supplies also have definite restrictions as to their application in regard to their input voltage with respect to their output voltage. The designer should consider these factors prior to the use of a nonisolated topology.

4.2.1 The Buck Regulator Topology

The buck regulator is the simplest of all the switching power supply topologies. It is also the easiest to understand and design. The buck regulator is also the most elementary forward-mode regulator and is the basic building block for all the forward-mode topologies. The buck regulator, though, exhibits the most severe destructive failure mode of all the configurations. For this reason, it should be used only with extreme discretion.

The buck regulator's basic operation can be seen as roughly analogous to a piston–flywheel combination. A steady-state DC current whose average value equals the output load current is always flowing through the inductor. The power switch, a power MOSFET in this case, acts only to replenish the energy in the inductor that was removed by the load during the MOSFETs off-time. The diode, called a *commutating diode*, maintains the flow of the load current through the inductor when the power switch is turned off. There are two current paths inside a buck regulator. When the power switch is conducting, the current is passed through the input source, the power switch, the inductor, and the load, after which it returns to the input source. Since the input source can provide much more energy than the load wants, the excess is stored in the inductor. When the power switch is off, the load current is passed through the commutation diode to the load and back again. The energy behind the sustained current flow is provided by the excess energy stored in the inductor, which is now being drawn on. This continues until the power switch is once again turned on and the cycle starts over again.

The voltage and current waveforms are shown in Figure 4.2. Analytically, they are quite easy to describe. First, the commutating diode's voltage is

$$V_d(Q_{on}) = V_{in} - V_{sat} \quad \text{(input voltage less the saturation drop of the power switch)}$$

$$V_d(Q_{off}) = -V_{fwd} \quad \text{(the forward voltage drop of the diode)}$$

The inductor's current can be seen as

$$Q_{on}: \quad I(\text{induct}) = I_{min} + \frac{[(V_{in} - V_{sat}) - V_{out}]\, T_{on}}{L}$$

$$Q_{off}: \quad I(\text{induct}) = I_{pk} - \frac{(V_{out} + V_{fwd})\, T_{off}}{L}$$

This yields a nice triangular current waveform whose edges are linear ramps. The inductor current is the sum of the power switch's and diode's current waveform. They are positive and negative ramps, respectively, riding on a current pedestal. The pedestal is indicative of the residual energy stored within the inductor acting as an energy reservoir. The residual energy is needed to quickly respond to changes in the load current before the control circuit can respond to the change. The DC average of this current waveform is equal to the DC current being drawn by the load.

Regulation of the output voltage is accomplished by varying the duty

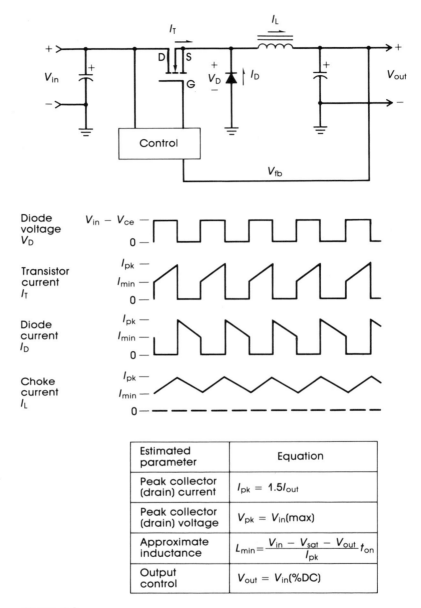

Figure 4.2
The buck regulator.

Estimated parameter	Equation
Peak collector (drain) current	$I_{pk} = 1.5 I_{out}$
Peak collector (drain) voltage	$V_{pk} = V_{in}(max)$
Approximate inductance	$L_{min} = \dfrac{V_{in} - V_{sat} - V_{out}}{I_{pk}} t_{on}$
Output control	$V_{out} = V_{in}(\%DC)$

cycle of the power switch's on-time versus off-time. This yields a control equation of

$$V_{out} = V_{in} \times (\text{duty cycle}) \quad [\text{approximation}]$$

As seen from the control equation, the higher the input voltage is above the output voltage, the narrower the on-time pulsewidth of the power switch. Conversely, the closer the input voltage gets to the output voltage, the more the duty cycle approaches 100 percent. It can also be seen that the output voltage is approximately the DC average of the power switch's output voltage waveform.

The buck regulator topology suffers from some limitations and problems imposed by the physics underlying its operation.

1. The input voltage must always be at least 1 to 2 V higher than the output voltage in order to maintain its regulated output. This can present a problem if the input supply could possibly approach the level of the output. This requirement is identical to the linear regulator where an input "headroom" voltage must be maintained for proper operation. As a result, the buck regulator can be used only as a step-down regulator.

2. When the power switch turns on, the diode is still conducting the inductor current. A diode takes a finite amount of time to assume a reverse-biased or off state, as specified by the reverse recovery time (T_{rr}) of the diode. While the diode is turning off, current will actually flow from the input line through the power switch and the diode to ground. This is actually an instantaneous short circuit across the input supply and adds stress to the power switch and diode. There is no way to eliminate this stress, but select the fastest reverse recovery diode possible (T_{rr}).

3. Semiconductor power transistors and MOSFETs almost always fail in the short-circuited condition when they do fail. This results in the input being short-circuited to the output load. Obviously, if there are no other means of protection, the output load circuitry would literally burn up. This is not a good way for a designer or a company to maintain a good reputation. The designer must add an overvoltage crowbar circuit to the output of the supply and a fuse in series with the input. The overvoltage crowbar [a silicon-controlled rectifier (SCR) driven by a voltage comparator] senses when the output voltage goes above a predetermined threshold, the SCR triggers, thus pulling an enormous current to the input ground return, which subsequently causes a series fuse to blow open. In reality the crowbar can be activated by spikes that may be asserted by the load or by a sluggish regulator in response to a rapid change in the load current. The regulator in this case enters a current foldback condition. This is an annoyance to the operator of the equipment, who must recycle (turn off and then turn on) the input power switch. The designer cannot ignore this failure mode. Component failures during the life of a product are a fact of life, so the designer should always create a design in anticipation of these events.

Although this topology is capable of delivering over 1000 W to a load in normal operation, it is not a popular choice among seasoned switching power supply engineers because of the above-mentioned shortcomings.

4.2.2 The Boost Regulator Topology

The boost regulator, otherwise known as a step-up regulator, is a flyback-mode topology. Its output voltage must always be higher than the input voltage.

The boost regulator uses the same number of components as the buck regulator, but they have been rearranged as seen in Figure 4.3. Its operation is also very much different from the forward-mode, buck converter. When the power switch is turned on the input voltage (V_{in}) is placed across the inductor. This causes the inductor current to linearly ramp up from 0 A until the power switch is turned off. During this time energy has been stored within the core material. At the instant the power switch is turned off the inductor voltage flies back above the input voltage. The inductor would reach an infinite voltage but is clamped to a value of the output voltage when the output rectifier becomes forward biased ($V_{out} + V_{diode}$). During the time which follows the energy within the core is emptied into the output filter capacitor and is made available to the load. This topology is limited to a 50 percent duty cycle since the core needs sufficient time to empty its energy into the output capacitor.

The mode of operation described above is referred to as the "discontinuous" mode of operation. This is the mode in which the vast majority of boost regulators operate. Its waveforms can be seen in Figure 4.3. The inductor voltage returns to zero (or V_{in} across the power switch) when the core has finished emptying its energy. The current ramp begins from zero. The other mode of operation is called the "continuous" mode. This occurs when the core cannot completely empty itself during the off-time of the power switch and some residual energy remains within the core. Now the inductor voltage does not return to zero and the current ramp rides on a pedestal that has a value proportional to the residual energy remaining in the core. Discontinuous-mode boost regulators can enter the continuous mode at low input voltages since the on-time pulsewidths grow larger in order to bring in the necessary energy required by the load. This does not allow enough time to empty the core's energy and usually indicates that the supply will soon be falling

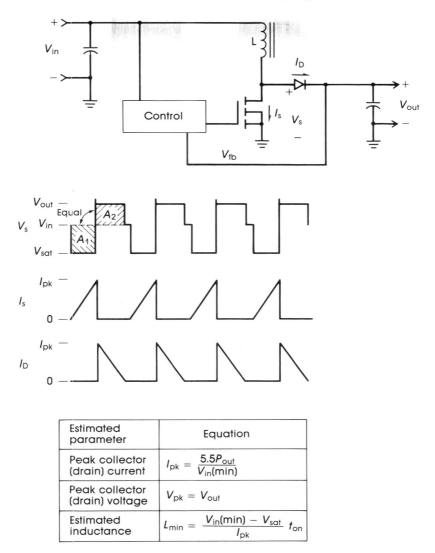

Figure 4.3
The boost regulator.

Estimated parameter	Equation
Peak collector (drain) current	$I_{pk} = \dfrac{5.5 P_{out}}{V_{in}(min)}$
Peak collector (drain) voltage	$V_{pk} = V_{out}$
Estimated inductance	$L_{min} = \dfrac{V_{in}(min) - V_{sat}}{I_{pk}} t_{on}$

out of regulation. The boost supply can be designed to operate in the continuous mode but this presents some stability problems, as described in Chapter 10.

An important question that must be answered during the design of the boost regulator is whether or not the inductor can provide enough energy to the load for its steady-state requirements. This can be determined by knowing the basic relationships within the boost regulator. The amount of energy stored within the core during each on period of the power switch is

$$W = \tfrac{1}{2} L(I_{pk} - I_{min})^2 \qquad (4.1a)$$

and the average power delivered to the output is

$$P_{out} = Wf \qquad (4.1b)$$

where P_{out} is the maximum output power capability of the inductor and f is the frequency of operation of the regulator. The P_{out} determined above should always be greater than the highest power needed by the load. If it is not, then the regulator will operate at light loads but will be unable to maintain regulation at the heavier loads.

So the problem is to make the inductance value low enough (but not too low as to resemble a short-circuit) to be able to accept sufficient energy at the lowest specified input voltage. This can be seen below.

$$I_{pk} = (V_{in} T_{on})/L \qquad (4.1c)$$

In order to maintain this energy, dictated by I_{pk}, at a low input voltage, the on-time must be increased. Soon a point is reached where the on-time pulsewidth extends into the period when the core is supposed to empty its energy into the output. Beyond this point any increase in pulsewidth only serves to add to the residual energy remaining in the core and the regulator will cease to regulate the output voltage. The designer's role is to determine the value of the inductance at which this occurs below the minimum specified input voltage.

This topology operates at about three times the peak current of forward-mode regulators. This is due mainly to having a 50 percent duty cycle limit. This high peak current limits its usefulness above 150 W since the stress on the semiconductor power switch becomes too great.

As with all non-transformer-isolated topologies, the ability of the boost regulator to prevent hazardous transients or failures within the supply from reaching the load is quite poor. For instance, if a large positive surge were to enter the regulator, it would exceed the output voltage and conduct directly into the load. Obviously, one could add transient protection, but many designers use the flyback regulator topology in place of the boost regulator. The transformer isolation vastly improves this condition.

4.2.3 The Buck–Boost Regulator Topology

The buck–boost regulator is a form of flyback-mode regulator, whose operation is very closely related to the boost regulator. It is also known

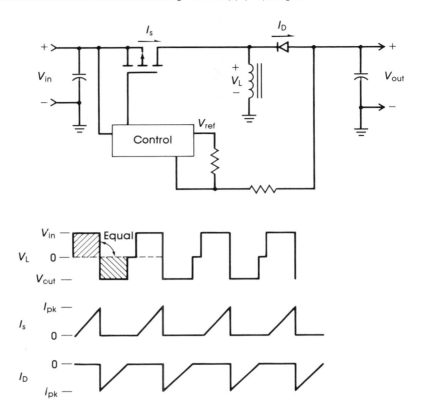

Figure 4.4

The buck–boost regulator.

Estimated parameter	Equation	Comment
Peak collector (drain) current	$I_{pk} = \dfrac{5.5P_{out}}{V_{in}(min)}$	
Peak collector (drain) voltage	$V_{pk} = -(V_{in}(max) - V_{out})$	V_{out} is negative
Estimated inductance	$L_{min} = \dfrac{V_{in} - V_{sat}}{I_{pk}}$	

as an *inverting regulator*. The difference between the boost and the buck–boost regulators, as seen in Figure 4.4, is that the positions of the power switch and the inductor have been reversed. Like the boost regulator, the inductor stores energy in the core material during the power switch's on-time. This stored energy is then released below ground (or the input return lead) through the rectifier into the output storage capacitor. The result is a negative voltage whose level is regulated by the duty cycle of the power switch. The buck–boost regulator is also limited to below a 50 percent power switch duty cycle since it requires time to empty the core of its stored energy.

The equations related to the core and its energy requirements are identical to those of the boost regulator. Once again, the inductor must store enough energy during each cycle of operation in order to sustain the load during that same period. This is determined at the low input voltage, where the voltage across the inductor is at its lowest and hence not able to absorb as much energy per microsecond, and at the maximum rated output load. This is the worst possible point of operation where the duty cycle approaches its physical maximum of 50 percent. As in all flyback-mode regulators, to increase the rate of storage within the inductor, if the energy is insufficient as seen by the regulator falling out of regulation at low input voltages, the designer decreases the inductance of the inductor. This helps the flyback-mode regulators operate at lower input voltages, but in this situation the peak currents can become too large to ensure reliable operation of the semiconductors.

The buck–boost regulator can suffer from catastrophic failure modes similar to those of either the buck or the boost regulator separately. First, if a negative transient is allowed to enter the regulator, a bipolar power transistor power switch may avalanche (overvoltage breakdown) the reverse-biased base–collector junction, which would cause the transistor to fail. This would then allow the negative transient voltage to enter the output and place an overvoltage stress on the load. Conversely, if a large positive transient enters the regulator, any semiconductor power switch will eventually enter avalanche breakdown and once again fail, short-circuit, and subsequently cause the rectifier to enter avalanche. This would then cause the positive input voltage to be placed across circuitry that is expecting only a negative voltage. Obviously, this will cause the load to fail. The good news is that for the more common source of power switch failure, overdissipation, the rectifier does offer some means of protection by virtue of its reverse-biased condition during the times when positive voltage is on its cathode. There is a very simple solution to these failure modes: the addition of a large overvoltage zener diode across the output capacitor and a fuse or circuit breaker ahead of the supply. For negative transient breakdowns, the zener will clamp the output voltage to safe limits and will blow the fuse. For positive transient breakdowns, the zener will act like a large forward-biased diode and shunt the current to ground and blow the fuse.

As in all power supply designs, it is better to suppress any transient before it enters the supply.

Once again, this topology is not typically used by experienced power supply designers because semiconductors offer poor isolation and protection against failures and failure-inducing conditions. This topology can reasonably be used only when a transformer or transformer-isolated supply is placed between the regulator's input and the input power source. An example of where one may use the buck–boost regulator would be as a board-level local regulator within a system that has a main power supply.

4.3 Transformer-Isolated Switching Power Supply Topologies

As one may have seen in the non-transformer-isolated regulator topologies, only the semiconductors provide the DC isolation from the input to the output. Semiconductors have relatively low breakdown voltages and exhibit the worst mean time between failures (MTBF) of all the components within any given power supply. This is not because they were manufactured incorrectly but because of heat factors and sporadic adverse operating conditions such as transients. The transformer-isolated switching power supply topologies rely on a physical dielectric barrier provided by wire insulation and/or insulated tape. The energy passes through a nonconducting ferrite material prior to reaching the output. This transformer isolation can withstand many thousands of volts before it fails and does provide a second dielectric barrier in the event of a semiconductor failure. This greatly discourages the domino effect of failures once a failure does occur within the final product.

On close inspection, the transformer-isolated regulators operate analogously to the nonisolated regulators. Similarly, there are forward- and flyback-mode regulators as in the non-transformer-isolated topologies. The transformer now provides a step-up/step-down function within the supply. The transformer also provides a second great advantage, the ease of adding multiple outputs to the power supply without adding additional separate regulators for each output. All these factors make the transformer-isolated regulator topologies an attractive choice for virtually all applications.

4.3.1 The Flyback Regulator Topology

The flyback regulator topology is the only flyback-mode regulator within the transformer-isolated family of regulators. It also is the simplest (contains the least parts) of all the transformer-isolated regulators. It is very closely related to the boost regulator but exhibits a great many advantages over its nonisolated counterpart. These advantages are so significant that it is selected over the boost in the vast majority of the applications.

As one may notice in Figure 4.5, the flyback design strongly resembles that of the boost regulator design except for the addition of a secondary winding on the inductor. Indeed, the size of the flyback regulator is only slightly larger than that of a boost regulator. It is the addition of this secondary winding that gives the flyback regulator its versatility. The advantages of the flyback over the boost or the buck–boost regulators are as follows.

1. More than one output is possible on one supply.
2. These outputs can be positive or negative in voltage.
3. The output voltage levels are independent of the input voltage.
4. The input voltage exhibits high dielectric isolation from the input to the output.

The flyback regulator actually works as a boost and a buck–boost regulator combined, and the input voltage can traverse the levels of any of the output voltages without affecting the operation of the supply.

The operation of the flyback regulator can be discussed by breaking one period of operation into two parts: the power switch's on-time and off-time. During the on-time, the full input voltage is placed across the primary winding of the transformer. This results in an increasing linear current ramp through the primary whose slope is $+V_{in}/L_{pri}$. This continues until the power switch is turned off. At this point the voltage as measured across the power switch (a MOSFET in this case) flies back to a voltage equal to the sum of the input voltage plus the turns ratio multiplied by the output voltage (plus a diode drop). So for example, if the transformer had a 1 : 1 turns ratio with a 5-V output, the flyback voltage would be 6 V above the input voltage (5 V + 1 V for the diode). During the flyback period (the power switch off-time) the output rectifier conducts, thereby passing the stored energy within the core material to the output capacitor and the load. This flyback period continues until either the core is depleted of energy, after which the power switch's voltage returns to the voltage of the input, or the power switch is once again turned on. The secondary current during the flyback period is a declining linear ramp with a slope of $-V_{out}/L_{sec}$. As one can see, since input and output voltages are rarely, if ever, equal and the primary and secondary turns may not be equal, the power switch on-time and the

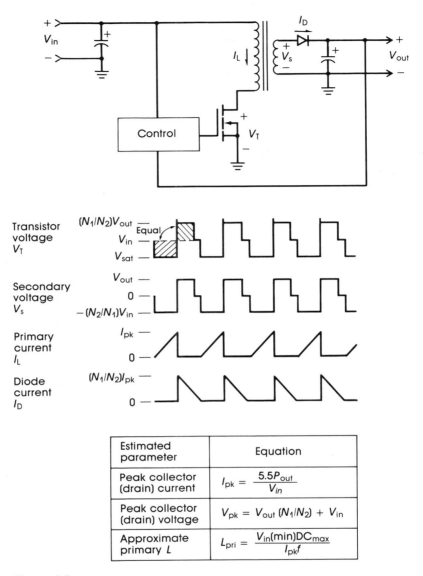

Figure 4.5

The flyback regulator.

flyback periods are rarely equal in time. One nice relationship is valid, though, when viewed from any one winding: the volt–time product of a winding during the power switch's on-time is equal to the volt–time product during the flyback period (see Fig. 4.5).

The flyback regulator can operate in either the discontinuous or continuous mode. In the discontinuous mode, the energy stored in the core

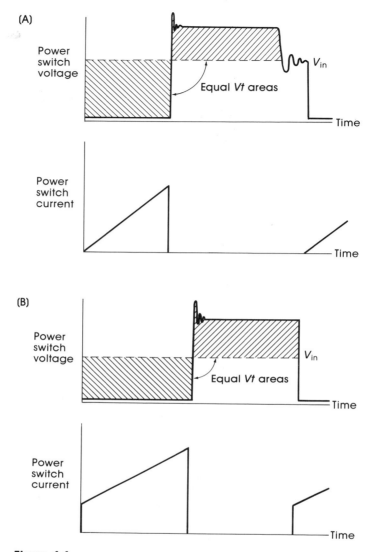

Figure 4.6

(A) Flyback operating in the discontinuous mode; (B) flyback operating in the continuous mode.

from the power switch's conduction period is completely emptied from the core during the flyback period. The occurrence of this mode can be seen easily by viewing the power switch's voltage and determining whether the flyback voltage returns to the input voltage level prior to turning back on again. There may be some ringing during this time since both the power switch and the diode are turned off, leaving the transformer completely unloaded. In the continuous mode, the power switch is turned on before the core empties itself of flyback energy (see Fig. 4.6). A typical flyback regulator may operate in either mode depending on the output load and the input voltage level. The flyback regulator will enter the continuous mode at low input line conditions when the increase in on-time pulsewidth does not allow enough time for the core to empty itself of the stored energy. For most flyback regulators, this also indicates that the regulator will soon be falling out of regulation. If the engineer designs the transformer for the heaviest expected load at the lowest expected input voltage, then everywhere else within the operating range the flyback regulator will simply shut down between cycles (discontinuous mode) and wait for the load demand to catch up with the power delivering capability. This capability allows the flyback supply to exhibit the broadest dynamic range (i.e., to regulate over a large range of input voltage and load current) of all the regulator topologies.

The operation of the flyback supply is a little more complicated to understand than that of the forward-mode regulators, but mathematically it is quite simple. Unlike a forward-mode transformer, the primary and secondary windings are wired out of phase so the primary and secondary currents do not flow simultaneously. Thus, the primary and secondary windings can be viewed as elementary inductors during their respective conduction periods. So, the input current can be described as follows:

$$i_{\mathrm{pri}} = \int_{t=0}^{T_{\mathrm{on}}} \frac{V_{\mathrm{in}}}{L_{\mathrm{pri}}}\, dt \qquad V_{\mathrm{in}} \text{ is a constant} \qquad (4.2)$$

or

$$i_{\mathrm{pri}} = \frac{V_{\mathrm{in}}(T_{\mathrm{on}})}{L_{\mathrm{pri}}}$$

Similarly, the secondary current is

$$i_{\mathrm{sec}} = \frac{V_{\mathrm{out}}(T_{\mathrm{flbk}})}{L_{\mathrm{sec}}} \qquad (4.3)$$

In the case of the secondary, the output inductance is a "charged" inductor discharging into a constant voltage load. Indeed, the output inductance may appear to be acting like a voltage source, but it is actually a current source being clamped by the voltage of the output capacitor. The energy entering the primary winding is given by

$$W = L \int_{t=0}^{T_{\mathrm{on}}} (i_{\mathrm{pk}} - i_{\mathrm{min}})\, dt \qquad \text{or} \qquad W = \tfrac{1}{2}L(i_{\mathrm{pk}} - i_{\mathrm{min}})^2 \qquad (4.4)$$

This represents the energy entering the core during each cycle of the regulator. To compare this energy to the demands of the load, the designer multiplies W by the operating frequency of the supply. The result is then given in watts, which can be directly compared to the demand of the load, which is also measured in watts. One can see from these equations the overwhelming temptation facing the designer of the flyback supply—the designer can deliver more power to the load and shrink the transformer by decreasing the primary inductance and accepting a higher peak current. This trade-off works fine to a point. At the higher peak currents, the reliability of the semiconductors within the regulator are adversely affected. So don't get too carried away with minimizing the size of the transformer.

The flyback transformer, because of its unipolar use of the $B–H$ curve, does exhibit very high flux excursions that could easily result in the core material entering saturation. When this happens, the linear current ramp exhibited by the flyback during the power switch on-time quickly becomes nonlinear and proceeds rapidly toward infinite current. This happens because the permeability of the core material in the saturation region quickly drops, which causes the value of the inductance to disappear, hence resulting in only the power switch appearing across the input line. Obviously, the power switch was not designed to endure such conditions. A problem can arise when the regulator is in operation at the high input line voltage and an instantaneous increase in the demand of the load occurs which causes the error amplifier to demand the widest pulsewidth possible. If the deadtime has been set for low input line conditions, the core could enter saturation. Within microseconds the power switch could fail. To avoid situations like this, the designer should place an airgap within the core to discourage the core from entering saturation (see Chapter 6).

4.3.2 The Push–Pull Regulator Topology

The push–pull topology is a transformer-isolated forward-mode regulator. Because it is a forward-mode regulator, it has the "buck" style $L–C$ filter network on its output(s). The transformer is used, in this case, as a stepping-up or stepping-down function of the chopped input voltage waveform before it is presented to the output $L–C$ filter(s). Unlike the flyback transformer, the push–pull transformer does not store any energy and the output current is drawn when either power switch is conducting. The push–pull topology utilizes a center-tapped primary winding. The input line is connected to the center-tap and a power switch is connected to both ends of the winding. The secondary voltage is full-wave-rectified and then presented to the output $L-C$ filter.

The push–pull regulator is what is called a *double-ended topology,* where two power switches share switching function. The power switches do not simultaneously conduct but alternate back and forth on alternate cycles. The two sides of the primary are wound in the same sense (or direction) but the current flows in the opposite direction. This results in the flux generated within the core material being driven in both the positive and negative flux polarities. This utilizes the core material in a more efficient manner, which can reduce the required core size. This can make the push–pull transformer core smaller than a comparable single-ended core if one neglects doubling of the windings on the transformer. The second advantage is that this topology can provide twice the output power of a single-ended topology operating at the same frequency. The two power switches share in the responsibility of eliminating the heat that is built up in them. This feature renders the push–pull topology capable of generating many hundreds of watts on its outputs.

The operation of the push–pull regulator is not difficult to understand (see Fig. 4.7). Only one transistor can turn on at any one time. When the transistor turns on, current begins to flow through its side of the primary. Simultaneously, one-half of the center-tapped secondary winding begins to conduct, thus forward-biasing its respective rectifier. This current then flows into the $L-C$ filter and is, in turn, stored by the inductor and capacitor. The voltage that appears on the $L-C$ filter has the peak value of the input voltage times the turns ratio from the primary to the secondary. This continues until the transistor is turned off by the controller. Next there must be a mandatory "deadtime" where neither transistor is conducting. This is because it takes a finite amount of time for the power switches to actually stop conducting current. In the case of bipolar power transistors, this could take as much as 2 μsec, depending on their drive circuits. For power MOSFETs, the time is much shorter, typically 50 to 400 nsec. It is critical that the power switches not conduct at the same time since this causes an effective short-circuited turn within the transformer, and astronomical levels of current could flow through the power switches, causing their immediate or latent destruction. When the opposing power switch turns on, a voltage equal to twice the input voltage appears across the inactive power switch. This is caused by current flowing in the opposite direction in the active half of the primary winding. Also during this period, the oppos-

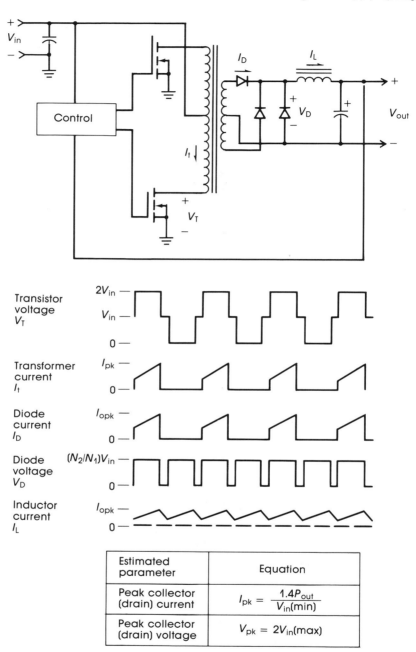

Estimated parameter	Equation
Peak collector (drain) current	$I_{pk} = \dfrac{1.4 P_{out}}{V_{in}(min)}$
Peak collector (drain) voltage	$V_{pk} = 2 V_{in}(max)$

Figure 4.7

Push–pull regulator.

ing secondary rectifier conducts. So each power switch and rectifier operates at half the power supply operating frequency, but the switching characteristics must be as good as if they were operating at the full operating frequency.

The current waveform as viewed from the primary of the transformer is not really a function of the transformer but rather that of the output filter section magnetically reflected through the transformer. The transformer, though, must have sufficient inductance and core cross-sectional area as to not enter a saturated condition. At the point of transformer saturation, all ability to couple energy to the secondary disappears and the energy exits the transformer through the power switches, resulting in their destruction. The secondary of the transformer, resembles a voltage source like the power switch in the buck regulator. Once again, the current waveform flowing through the output filter inductor follows the buck regulator mathematical model:

$$I(\text{induct}) = I_{\min} + \frac{[(V_{\sec} - V_{\text{rect}}) - V_{\text{out}}] T_{\text{on}}}{L}$$

(on)

$$I(\text{induct}) = I_{\text{pk}} - \frac{(V_{\text{out}} + V_{\text{rect}}) T_{\text{off}}}{L}$$

(off)

Like the buck regulator, the inductor must not become emptied of flux. Typically, the minimum current point is 50 percent of the rated load current. So the secondary current waveform is a current ramp sitting atop a pedestal of approximately one-half of the output current. This waveform is then multiplied by the inverse turns ratio of the transformer to obtain the waveform as seen from the primary.

Although this regulator topology can handle up to several kilowatts in output power, it suffers from one serious flaw. Its problem arises from certain definite real-world factors. Specifically, no two power switches are identical and no two halves of a center-tapped winding are identical. This means that one side of the primary will have a fraction of a turn less than the other side or that the power switch will turn off slightly slower or have a slightly lower saturation voltage. This condition guarantees that the transformer core will never operate symmetrically about the origin of the $B-H$ curve. This results in one side of the primary having a higher peak current and being nearer to saturation than the opposing side. This, in itself, is not a problem until a step increase output load current occurs. The error amplifier will drive the power switches to their maximum pulsewidth, causing the higher current side to enter saturation, which may result in destruction of that power switch. This is called *core imbalance*. The only way to avoid this situation is to add a high-speed current-sensing and cutoff circuit to the control section. Since normal operational amplifiers cannot respond that quickly, this circuitry usually must be built from discrete components. This can add significantly to the cost of the supply and makes the push–pull topology unattractive for most designs. Current-mode control may reduce the problem due to core imbalance since it senses the instantaneous primary currents, but it can remain a major problem. The more experienced power supply designers prefer to use the half- or full-bridge topologies in place of a push–pull topology.

4.3.3 The Half-Bridge Regulator Topology

The half-bridge regulator is another form of a transformer- isolated forward converter. As one can see in Figure 4.8, the arrangement of the components around the primary circuit is drastically different from that in the push–pull regulator. The half-bridge regulator has only one primary winding, which is connected between a pull-up/pull down arrangement of power switches (similar to a totem-pole driver) and the center node between two series capacitors wired between the input voltage and ground. Like the push–pull regulator, it operates the transformer core in the bipolar flux mode of operation. The capacitor center node voltage sits at approximately one-half of the input voltage, and the power switches present the other end of the primary winding with an alternating input voltage and ground signal. This means that only half the input voltage appears across the primary winding. This results in an average and hence peak current twice that of the push–pull regulator with an identical output power. Thus, the half-bridge regulator is not as suitable for very high power operation as the push–pull regulator but has one overwhelmingly positive feature: it exhibits an intrinsic self-core balancing. The core balance is accomplished by the capacitors. The center node voltage will adjust in the direction of higher flux density within the transformer. This reduces the voltage across the primary in the direction of impending saturation, which, in turn, centers the $B–H$ excursions within the transformer, eliminating the need for an expensive high-speed overcurrent-sensing circuit.

The operation of the secondary and output filter circuits is identical to that of the push–pull regulator. The transformer turns ratio is different since only half the input voltage appears across the primary winding.

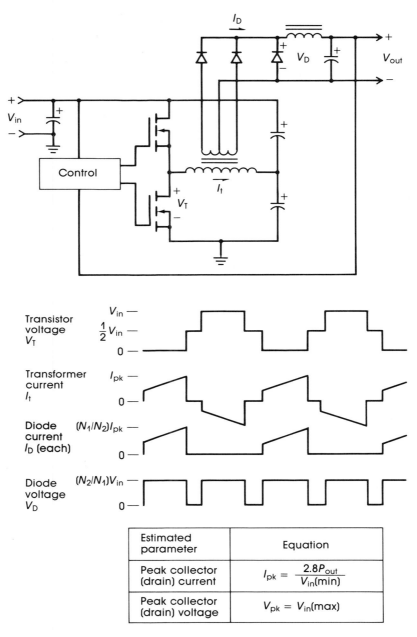

Estimated parameter	Equation
Peak collector (drain) current	$I_{pk} = \dfrac{2.8 P_{out}}{V_{in}(min)}$
Peak collector (drain) voltage	$V_{pk} = V_{in}(max)$

Figure 4.8

Half-bridge regulator.

One major difficulty encountered in using the half-bridge regulator topology is how to drive the upper power switch whose emitter or source is riding on a high-voltage AC waveform. Its drive signal must be referenced to this AC waveform. The common method is to drive the upper power switch with an isolated pulse transformer and reference the secondary to the AC waveform. This adds cost to the power supply, but this cost can be moderated by adding a second output winding to the drive transformer and driving the lower power switch with the same transformer. This eliminates the need for a ground-referenced driver circuit and a floating driver circuit for both power switches. It also allows the controller section to be transformer-isolated from the input line, if desired.

Since only half the input voltage appears across the primary winding, the peak current in this topology is twice as high as in the push–pull topology. This means that power switches reach their maximum ratings at half the output power of the push–pull regulator. The half-bridge regulator topology is used in applications requiring between 150 and 500 W of output power. Below 150 W the flyback regulator is more cost-effective, whereas above 500 W the power switch's reliable operation is questionable. The core-balancing feature of the half-bridge regulator nonetheless makes it the popular choice within this power range.

4.3.4 The Full-Bridge Regulator Configuration

The full-bridge converter is the last of the popular transformer-isolated pulsewidth-modulated (PWM) topologies. Like the other double-ended regulators, its transformer's flux is driven in both the positive and negative polarities. Its performance with respect to output power is significantly improved over that of the half-bridge converter. This is because the balancing capacitors are replaced with another pair of half-bridge-style power switches identical to the first pair. This time two of the four power switches are turned on simultaneously. During one conduction cycle either (1) the upper left and the lower right power switches or (2) the upper right and lower left power switches are turned on. Each associated pair of power switches conduct on alternate cycles. This places the full input voltage across the primary winding, thus reducing the peak currents in the primary for any output power compared to the half-bridge regulator. This effectively doubles the maximum power-handling capability of this topology over the half-bridge (see Fig. 4.9).

Once again, the problem of driving the upper power switches is present in the full-bridge topology. The cost of the drive transformer be-

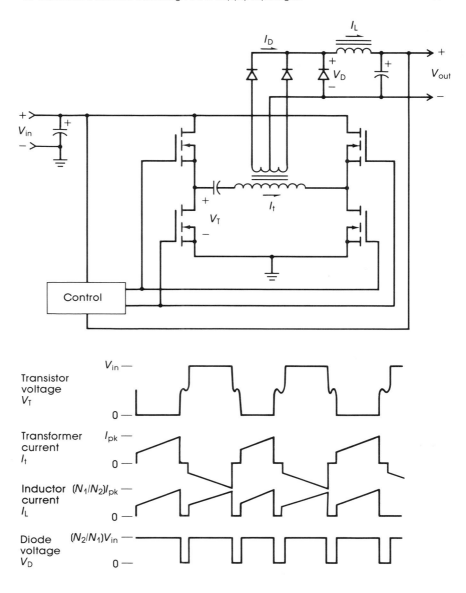

Estimated parameter	Equation
Peak collector (drain) current	$I_{pk} = \dfrac{1.4 P_{out}}{V_{in}(min)}$
Peak collector (drain) voltage	$V_{pk} = V_{in}$

Figure 4.9
Full-bridge regulator.

comes minor when delivering such a high power to the load. Also, since the power switches are driven in pairs, the designer needs only to add two more secondaries to the half-bridge drive transformer to accommodate the added pair of power switches. The control circuitry remains unchanged.

Core balancing is achieved by placing a small nonpolarized capacitor in series with the primary winding. The average DC voltage across the capacitor reduces the voltage across the primary winding in the direction of impending saturation. Core saturation at the power levels at which the full-bridge regulator operates would mean instant destruction of the conducting power switches. It is a good idea to design the transformer for slightly lower maximum flux excursions than normal simply for this reason.

The full-bridge regulator topology is used in applications requiring output powers of 300 W to many kilowatts.

5

Semiconductors Used in a Switching Power Supply

The discovery and subsequent low-cost availability of the semiconductor marked a turning point in the history of the practical switching power supply. The concept had been known since the 1930s, but few power supplies had been built using switching techniques. The semiconductor afforded the designer the advantages of size, weight, speed, and ease of use. Switching power supply technology advanced with the latest advances in semiconductor technology. Today, semiconductors have reached a high level of sophistication in regard to their applicability to switching power supplies. Even with this high level of sophistication, the semiconductor components are still the most fragile elements within the switching power supply. So to create a reliable supply design, the designer must have a good understanding of not only how to use the parts but also their causes of destruction. The application of the semiconductors also has a great bearing on the supply's efficiency.

5.1 Bipolar Power Transistors

Bipolar power transistors have been used as power switches within switching power supplies since the advent of semiconductors. They are still used in a great percentage of the commercially marketed switching power supplies. Power MOSFETs are continuing to displace many bipolar power transistors in their role of power switches as MOSFET technology improves. The transistor, though, has a very valid place in the world of switching power supplies.

Bipolar power transistors are current-driven devices. That is, it requires a current into the base to cause a collector current to flow or turn the transistor on. When the transistor is used in the linear mode, the relationship of base current to collector current is called the *current gain* of the transistor, which is described as

$$I_B = \frac{I_c}{h_{FE}} \qquad (5.1)$$

The transistor when used in switching power supplies, though, is driven into saturation or near saturation in order to minimize the collector-to-emitter voltage and maximize the collector current. That is to make it appear as close to an ideal closed switch as possible. This makes the relationship become

$$I_B > \frac{I_c}{h_{FE}} \qquad (5.2)$$

This means that the designer must overdrive the base with enough current to guarantee saturation or near-saturation of the collector-to-emitter voltage in order to minimize the conduction losses within the transistor. Therefore, the designer must always do a worst-case design for the minimum specified gain of the selected transistor and drive the base with sufficient base current in order to saturate the transistor during the highest anticipated peak collector current. This highest anticipated collector current is dictated by the output filter inductances, the input voltage, the maximum conduction period, and the load. This is the typical approach to driving power transistors and it is called *fixed base drive*. This relatively high base current can present a problem to the designer—namely, where to draw the current from. If it is drawn from the input line, the base drive loss can be quite high ($V_{in} \cdot I_b$). Many novel approaches have been used to reduce this loss, such as adding a winding to the transformer that provides a much lower voltage than the input voltage. This reduces the power loss in deriving the base current from a much higher voltage. The transistor will also turn off more slowly in proportion to the degree of overdrive. In a nonoptimized supply, the fixed base drive losses can run as high as 40 percent of the total supply losses. Another approach, called *proportional base drive,* uses a base drive transformer that sums in some of the collector current into the base drive current. This provides only enough current to the base to maintain saturation of the transistor at that instant in time. Since this added current is derived from a low-voltage, current-fed winding on the transformer, the losses are quite low and the turn-off times are fast. So the consideration of how to drive the transistor has a large impact on the supply efficiency (see Fig. 5.1).

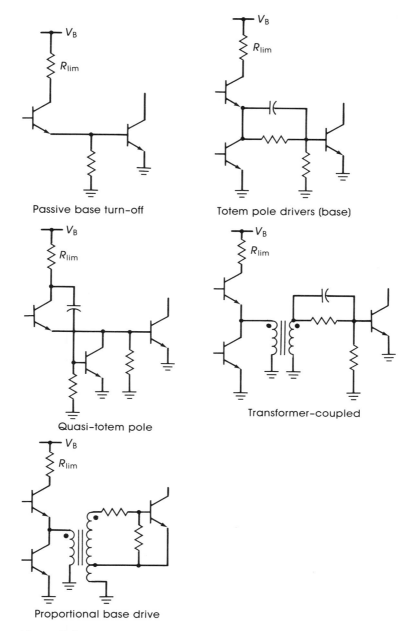

Passive base turn-off

Totem pole drivers (base)

Quasi-totem pole

Transformer-coupled

Proportional base drive

Figure 5.1
Bipolar transistor base drive circuits.

Excess base current

From fixed base drive circuit

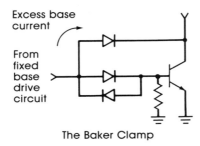

Figure 5.2

Nonsaturated base drive circuits.

The Baker Clamp

From low–current fixed base drive circuit

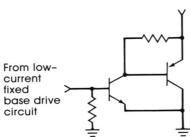

The equivalent *NPN* configuration

The bipolar power transistor switches much slower than a power MOSFET. Its switching speed is also greatly affected by the method of driving the base (see Fig. 5.2). In fixed base drive designs, the typical switching speed can range from 100 nsec to 1.5 μsec. In switching power supplies, the switching speed has a direct affect on the transistor's "switching losses." Switching losses are caused by the finite period of time for the collector-to-emitter voltage to change from a saturated "on" level to an "off" level and vice versa. During this time there is still collector current flowing as a result of the inductive load. This instantaneous product of V_{ce} and I_c can amount to tens or hundreds of watts instantaneously within the transistor, and the average loss over time is directly proportional to the supply's operating frequency. During turn-on and turn-off, the base is better viewed as a small capacitor from the base to the emitter. The rate by which this capacitor is charged or discharged dictates how rapidly the transistor will switch between cutoff and saturation. This makes the transistor drive circuit a little difficult to design. During the "on" period the drive circuit wants to look like a current source, but during the transitional periods it wants to look like a switched voltage source. The optimum driver topology is one that has an active pull-up and active pull-down such as a totem-pole configuration. On-time base drive current limiting should be split between two resistors, one in the supply side of the pull-up driver and one in series between the totem-pole and the base of the power transistor. A small capacitor on the order of 50 to 200 pF is placed in parallel with the base resistor. This "speed-up" capacitor, as it is called, stores just enough charge to quickly push and pull charge into and out of the base–emitter capacitance of the power transistor. The speed-up capacitor actually presents a negative voltage spike to the base during turn-off and reduces the effects of the current crowding failure mode of the transistor. Although the base–emitter voltage likes to be driven slightly negative during turn-off, the designer must not drive the base too negative or the base–emitter junction will break down (overvoltage avalanche). A good rule of thumb value is -5 V maximum. In transformer-coupled and proportional base drive methods, many of the same items apply. A good, tightly wound base drive transformer couples the characteristics of the drivers through to the power transistor. So if a passive turn-off method were used on the primary of the drive transformer, the transistor would still turn off slowly. If a totem-pole style driver were employed, the power transistor would switch quickly.

Certain key transistor specifications that should be discussed and observed during the design process are rise and fall times and storage time. The values given in a typical data sheet are derived specifically from the test setup shown in the data sheet. The actual values within your power supply are completely dependent on the drive configuration that is used. Since the test setup does not change significantly from transistor to transistor, it does give the designer an idea as to which transistor switches faster. The rise and fall times are controlled by how rapidly the drive circuitry can charge and discharge the base–emitter capacitance. These times occur when the peak switching losses occur within the transistor and have a direct bearing on the supply's efficiency. The storage time is controlled by how much overdrive the base receives just prior to turn-off. This is the period of time between when the base drive is removed and the collector-to-emitter voltage just begins to change. Although this period does not appreciably add to the losses within the transistor, it does give the supply a fixed minimum pulsewidth and deadtime that can occur within the power supply.

Two major failure modes of the bipolar power transistor are avalanche breakdown when the V_{ce}(sus) parameter is exceeded and second breakdown phenomenon (see Fig. 5.3). Avalanche breakdown occurs when the transistor is in the cutoff state and a voltage surge or spike is allowed to reach the collector terminal. The more common and more difficult to understand phenomena are the second breakdown and the current-crowding phenomena. These occur during the period of turn-on and turn-off, respectively. These are voltage-dependent phenomena where

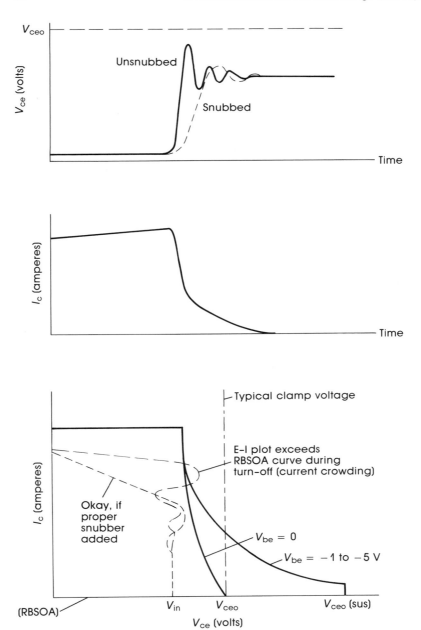

Figure 5.3

What the SOA curve means to the real world: Example of a current–crowding failure condition.

there is a current flowing through the die with a relatively large voltage from collector to emitter. This equates to a large instantaneous power dissipation that is not uniformly distributed across the surface of the die. At turn-on, the base–collector voltage gradient causes the collector-to-emitter current flow to concentrate below the emitter finger edges, and at turn-off the current flow concentrates below the center of the emitter

Figure 5.4

Physics behind second breakdown.

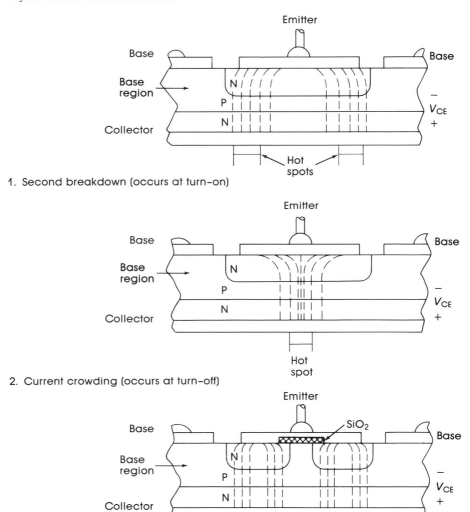

1. Second breakdown (occurs at turn-on)

2. Current crowding (occurs at turn-off)

3. Solution: Hollow emitter distributes current concentrations more evenly over the die. (Switchmode III and IV)

fingers. Both situations exhibit current concentrations flowing along a vertical path where there is a significant voltage drop from end to end. These result in occurrences of some of the highest instantaneous power that the transistor sees in its operation, concentrated over a small portion of its die area. It also does not occur uniformly over the length of the emitter finger. Second breakdown failures (see Fig. 5.4) typically occur close to the base bonding pad, and current-crowding failure typically occurs at the opposite side of the die from the base bonding pad. It is also a puzzling failure to identify since it can happen instantaneously even when the transistor is cool to the touch. This condition is inevitable, so to minimize this instantaneous power, one must switch the transistor as fast as possible, which reduces the time period for the instantaneous $E \times I$ product. This is done by two methods. The first is to operate the transistor just below saturation. This reduces the base storage time and fall time of the transistor, which is dependent on how far the transistor is driven into saturation. The second method, if one is driving the transistor into saturation, is to pull the base–emitter voltage slightly negative at the turn-off transitions. Some guidelines during the transistor selection process are

1. Use a transistor specified for switching applications, such as the Motorola Switchmode series. These have a hollow emitter construction that tends to distribute the current concentrations more evenly across the die, thus reducing any "hot-spot" temperatures.
2. Select a transistor whose breakdown voltage is 25 percent higher than the highest expected collector-to-emitter voltage.

Using the bipolar power transistor as the power switch within a switching power supply design can be complicated, but many years of usage has yielded a large pool of design methods which accommodate the transistor's characteristics. So don't reinvent the wheel.

5.2 Power MOSFETs

Power MOSFETs are quickly gaining popularity within the switching power supply field as fast power switches. Power MOSFET technology has matured greatly in the recent past and exceeded the performance of the bipolar power transistors. Power MOSFETs now switch approximately 10 times faster than their bipolar transistor counterparts when driven by fixed base drive methods. MOSFETs have also attained saturation voltages very comparable to those of the power transistors. This makes the power MOSFET the better choice for switching power supplies in the majority of applications.

The power MOSFET is an isolated-gate, voltage-driven device. That means that it takes much less average current to drive the gate of a MOSFET. The gate drive voltage, though, must reach 10 V for most MOSFETs to attain a saturated drain-to-source voltage. The gate of a MOSFET can often be viewed as a capacitor from gate to source whose value can range from 900 to 2000 pF. So in a DC application a current flow in only the nanoampere range is needed to maintain the on or off state, but switching the MOSFET quickly between the on and off states may require an ampere or more of peak current. This means that the driver should be a low-impedance active pull-up/pull-down type driver, such as a totem-pole driver (see Fig. 5.5). The totem-pole driver should also have a solid, well-bypassed voltage source as its supply in

Figure 5.5
Power MOSFET gate drive circuits.

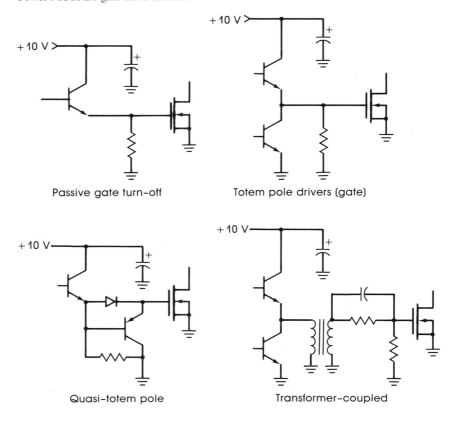

Passive gate turn-off   Totem pole drivers (gate)

Quasi-totem pole   Transformer-coupled

order to source and sink the relatively high peak currents. The output of the driver should only resemble a switched voltage source to the gate of the MOSFET. With an active pull-up/pull-down driver one can easily attain switching speeds of 30 to 50 nsec without much trouble. In some cases, though, it may be necessary to decrease the switching speed. To do this, it is often recommended to add a current-limiting resistor directly in series with the gate. This approach will give the designer better control over the actual switching speed. It is not recommended that the switching period be made longer than 1 μsec since the added switching losses may become excessive at the supply's operating frequency and contribute to a thermal problem within the MOSFET.

The physical construction of the MOSFET makes it different in its application within switching power supplies. One relief for the designer is that the power MOSFET does not suffer from second breakdown and the current-crowding phenomenon as does the bipolar power transistor. Although the switching loss still heats up the MOSFET die, the physical concentration of current during turn-on and turn-off is eliminated. This gives the power MOSFET a square reverse-bias safe operating area (RBSOA) and commutating safe operating area (CSOA) characteristic. One area of new concern for the designer is that the drive impedance should not exceed 200 Ω. This is needed not only for the high switching speed but also to swamp out the effects of the miller capacitance. The miller capacitance is a small 80- to 100-pF equivalent capacitor connected between the drain and the gate. Although it is a small value, it has a high voltage across it (many millicoulombs of stored charge). When the MOSFET is turned on or off, the miller capacitor couples the fast transition in drain voltage directly to the gate. In high-impedance drive circuits the MOSFET will actually appear to be oscillating when the MOSFET's threshold voltage is reached. Although it is not lethal to the MOSFET, it can add significantly to the losses within the MOSFET. Even in drive circuits with reasonable output impedances, the gate voltage waveform can be seen to have "plateaus" on the transition edges.

Although power MOSFETs are better suited for high-frequency switching power supplies than most bipolar transistors, their failure modes must be understood in order to design a reliable switching power supply. The following failure modes are listed in order of severity, that is, how often, and how quickly the failure occurs.

1. *Avalanche breakdown.* This failure occurs when the maximum breakdown voltage rating (V_{dss}) is exceeded during turning-off or turned-

off conditions. The major contributor to this problem is the leakage inductance of the primary winding of the transformer in typical power supplies, along with the use of slow rectifiers with slow forward recovery times (T_{frr}). This failure condition occurs when high voltage spikes during the turn-off transitions on the drain of the MOSFET. In nonruggedized MOSFETs the problem is caused by the parasitic bipolar transistor (which is part of the internal parasitic diode). Once avalanche is entered, the parasitic transistor may actually exhibit an avalanche rating of less than the MOSFET itself. This means that if the input voltage is greater than this transistor's avalanche voltage, the MOSFET is guaranteed to fail if avalanche were ever encountered. In ruggedized MOSFETs, the parasitic transistor has been optimized to have an avalanche voltage higher than the MOSFET rating. It is said that the improved parasitic transistor may even be used as a zener clamp. The avalanche specifications do have maximum energy limits, and it is not recommended to use the MOSFET in this manner. To reduce avalanche-inducing spikes, it is recommended to first rewind the transformer for tighter coupling (minimum leakage inductance) and to use faster recovery diodes on the secondary. If that still does not yield adequate results, the addition of snubbers and/or clamps may be necessary. A second major contributor is the occurrence of lightning strikes and voltage surges from the input power line. These will show up as field failures if the designer did not account for them in the design. One best attacks these by placing transient suppressors immediately behind the RFI filter choke at the input of the power supply.

2. *Commutation of the intrinsic diode.* Some manufacturers of power MOSFETs do not have the same current rating and speed rating of the diode when compared to the MOSFET itself. So when current is allowed to flow through the intrinsic diode, several things happen. First, it contributes to the power dissipation inside the FET, and it takes a long time for some diodes to turn off (see also Fig. 5.6). The designer must add the additional power dissipated in the diode to the losses of the MOSFET in order to properly represent the losses within the package for the thermal calculations. This is an easy factor to overlook, which can lead to overdissipation in the MOSFET. When the diode has a slow reverse recovery time, another FET within the circuit may turn on, causing very high reverse currents to flow through the diode. This can represent a virtual short circuit from the input supply to ground, thus instantaneously causing the MOS-

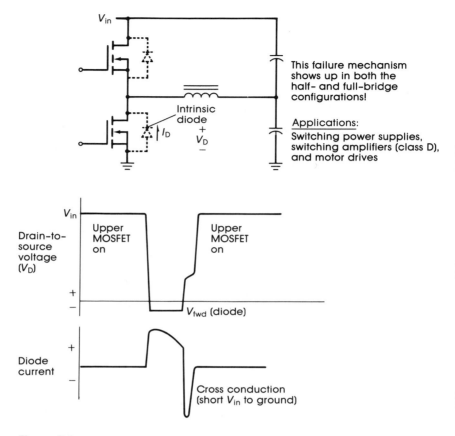

Figure 5.6

Commutating *dV/dt* failure condition.

FET to fail. This occurs predominantly in half- and full-bridge regulator configurations. The solution to this problem is to add two ultra-fast-recovery diodes to each MOSFET in the bridge. The first diode is directly in series with the drain lead and prevents current through the intrinsic MOSFET diode. The second diode is placed in parallel with the MOSFET and the first diode to replace the function of the intrinsic diode.

3. *High gate drive impedance.* If the gate drive impedance is too high, then the miller capacitor from the drain to the source can couple enough energy to the gate to cause the MOSFET to "bounce" on turn-on and turn-off. In other words, the FET will oscillate at each transition, thus making the switching loss very high, and overdissi-

pation may result. Gate driver impedances in switching applications
(>15 kHz) should be less than 200 Ω in total.

4. *Overdissipation.* This occurs when the designer does not consider all
the losses within the MOSFET. These should be measured with an
oscilloscope current probe and an oscilloscope voltage probe. This is
for performing a "real-time" graphical integration and the calculation of the instantaneous power being dissipated within the MOS-
FET. The losses should be the saturation losses (on time), the switching losses, and the diode conduction loss. A designer who ignores
any of these losses may face a problem later. Proper heatsink design
is also required. Remember, the specified power ratings on the devices assume an infinite heatsink!

Other types of MOSFETs are logic-level MOSFETs and current-
sensing MOSFETs. Logic-level MOSFETs are guaranteed to attain saturated output conditions at 5 V (V_{gs}) instead of 10 V as in the typical
MOSFET. This allows logic circuitry and other 5-V circuitry to switch
high currents. The one major drawback to their use in switching power
supplies, is that the gate-to-source capacitance is double that of typical
MOSFETs. This makes driving them more difficult.

Current-sensing MOSFETs (Fig. 5.7) incorporate a current mirror
inside the FET. The current passing through the mirror is exactly proportional to the current passing from the drain to the source but at a
much lower level. This enables the designer to determine the current
flowing through the MOSFET, but at a small fraction of the losses as-

Figure 5.7

Comparison of resistive current-sensing techniques.

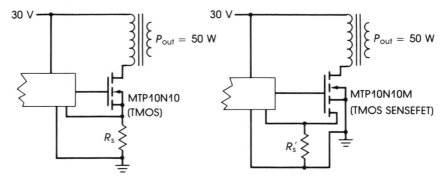

The SENSEFET essentially provides losses current sensing 1/1000 of the power loss.

sociated with current sensing on the drain or source. It also will cost less than the other methods. There are two terminals associated with this function: the mirror pin from which the mirrored current flows and a *kelvin contact*, which is a low-noise source connection. The mirror current is transformed into a voltage for the control circuitry by placing a small resistor between the mirror and the kelvin leads. The sense resistor should not exceed 200 Ω since the temperature coefficient of the MOSFET can lead to inaccuracies over the operating temperature range.

5.3 Rectifiers

Rectifiers play a critical role inside switching power supplies. They can also dissipate half the losses within a typical power supply. Since the rectifier is a two-lead device, there is not a whole lot that the designer can do other than select the best diode for the application.

The important parameters for rectifiers in regard to switching power supply design are (see Fig. 5.8):

1. *Forward voltage drop.* This is the voltage drop across the diode when a forward current is flowing in the diode. The lower the drop, the more efficient the operation.
2. *Reverse recovery time.* This is the amount of time it takes the diode to turn off after a forward voltage is removed. It takes a finite period of time for the diode current to stop after a large reverse voltage is applied to the terminals. This is an instantaneous power spike at the diode turn-off. Obviously, the shorter the T_{rr}, the less loss within the diode.
3. *Forward recovery time.* This parameter, which is not always specified by the diode manufacturers, is how long it takes the rectifier to begin to conduct forward current after a forward voltage is applied. The faster the turn-on time, the less spiking occurs when transformers and inductors are allowed to be unloaded while the diode is turning on.
4. *Reverse blocking voltage.* This is simply the amount of reverse voltage the rectifier can withstand before breaking down. When determining the necessary blocking voltage of the diode the designer should consider not only the waveforms but also any voltage spikes that might appear across the diode.

The four basic types of rectifiers are listed below in the order of most desirable parameters (see also Fig. 5.9).

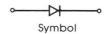

Symbol

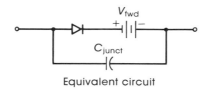

Equivalent circuit

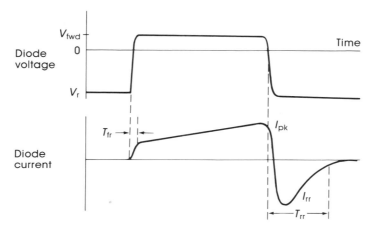

Figure 5.8

The rectifier.

Parameter	Schottky	Ultrafast	Fast recovery	Standard recovery
Forward voltage, V_{fwd}	0.5–0.6 V	0.9–1.0 V	1.2–1.4 V	1.2–1.4 V
Reverse recovery time, T_{rr}	< 10 nsec	25–100 nsec	150 nsec	1 μsec
T_{rr} form	Soft	Abrupt	Soft	Soft
DC blocking voltage, V_r	20–100 V	50–1000 V	50–1000 V	50–1000 V

	Low-voltage, high-current outputs	Higher-voltage, low- to high-current outputs	AC line only
	Diode clamps		
	Most efficient ←	→ Least efficient

Figure 5.9

A comparison of different rectifier technologies.

1. *Schottky barrier rectifier.* This type of diode has the lowest forward voltage drop of all the diodes and has no reverse recovery time. The only drawback is that it is only usable up to 100 V of reverse voltage and has a high reverse leakage current. This makes it appropriate for the low output voltage, higher current outputs.

2. *Ultra-fast-recovery rectifier.* This diode exhibits reasonable forward voltage drop typically (0.9 V) and has a fast recovery time ($\sim$ 35 to 50 nsec). These devices can also handle high reverse voltages up to 1000 V.

3. *Fast-recovery rectifier.* This type of diode exhibits a higher forward voltage drop and slower reverse recovery times than do the ultra-fast-recovery rectifiers (1.4 V and 200 nsec). They do cost less than the ultra-fast rectifiers, and if efficiency is not an issue, these rectifiers would be a good choice.

4. *Standard-recovery rectifiers.* These diodes are not intended for use as rectifiers within switching power supplies. They should only be used as 50–60-Hz rectifiers on the input lines of off-line switching regulators.

5.4 Switching Power Supply Control Integrated Circuits

In recent years, a large variety of ICs have emerged in the marketplace that have facilitated the implementation of higher-level functions within a switching power supply. The selection of the best controller IC should be done after the designer knows the level of functionality that is expected of the system. Although control ICs can have very different block diagrams, they have certain common circuit functions, including:

1. An oscillator that sets the basic frequency of operation of the supply and also generates a ramp waveform for use in voltage to PWM conversion.

2. Output drivers that provide enough drive current for low- and medium-power applications.

3. A voltage reference that provides the overall power supply "ideal" reference to which the output voltages are compared. It also can provide a stable voltage for other control functions.

4. A voltage error amplifier that performs the high gain voltage comparison between the output voltages and the stable reference.

5. An error voltage-to-pulsewidth converter that sets the duty cycle out-

put in response to the level of the error voltage from the voltage error amplifier.

These functional blocks form the basic PWM control IC. Other functions sometimes included in the IC provide some higher level of functionality is usually needed in a switching power supply, such as:

1. An overcurrent amplifier that protects the supply from abnormal overcurrent conditions within the load.
2. A soft-start circuit that, as the name implies, starts the power supply in a smooth fashion, reducing the inrush current exhibited by all switching power supplies during this period.
3. Deadtime control that fixes the maximum pulsewidth the control IC can generate, thus preventing the occurrence of simultaneous conduction of two power switches or 100 percent duty cycles.
4. Undervoltage lockout to prevent the supply from starting when there is insufficient voltage within the control circuit for driving the power switches into saturation.

To begin the selection process, the designer must first determine the topology of the supply that is needed for the application. This provides the first requirement of the controller IC, which is whether a single- or double-ended controller is needed. Single-ended controllers are those supplies that require only one power switch to implement the design. These are all the non-transformer-isolated topologies and the flyback topology. These ICs have only one output driver. Double-ended ICs are those that have two output drivers that would be applicable for the push–pull, half-bridge, and full-bridge topologies. These ICs include an additional feature called *double-pulse lockout,* which ensures that the same power switch cannot turn on consecutively, which could cause saturation of the transformer. A second factor is which type of power switch is to be used within the supply. Some control ICs have single-output transistors for their output drivers. These are better suited for bipolar power transistors. Almost always, additional drivers external to the IC need to be added to provide enough current drive to the transistor. For power MOSFETs, the ICs with totem-pole drivers are a better choice. These output drivers are ideal for sourcing and sinking the large peak currents needed to drive the capacitive gates of the MOSFETs. Although either type of output driver can drive either type of power switch, these choices provide a minimum of additional external components needed to implement the drive.

5.4.1 Voltage-Mode Control

Finally, the mode of control needs to be considered. These are basically single-loop control or multiloop control. Single-loop control is commonly called "voltage-mode" control. This is the traditional method of control where only the output voltage is sensed and compared to the reference in order to control the duty cycle of the power switches. Voltage-mode control provides adequate control for the power supply but can introduce problems in its transient load response time and line regulation. These types of controller have been the parts traditionally offered, such as the one shown in Figure 5.10. It compares the error to the ramp waveform created in the oscillator section to determine the pulsewidth for the power switch. This performs quite well when the loads are constant. But as one can see, only the output is sensed at the

Figure 5.10

Representative voltage-mode control IC (courtesy of Motorola Semiconductor, Inc.).

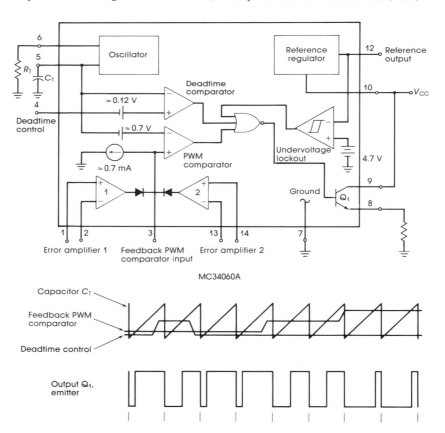

rear of the power supply. If the load or the line changes quickly, the magnetic elements and the filters provide a severe time delay before the overall supply can respond to these changes. For this reason alone, multiloop control was invented.

Some of the popular voltage-mode control ICs are listed below.

Single-Ended Controllers	Double-Ended Controllers
SG1524	SG1525/26/27
MC34060	TL494/495
uA78S40	
MC34063	
NE5560	

5.4.2 Current-Mode Control

Current-mode control now adds a second control loop to the voltage feedback loop. In place of the ramp from the oscillator, as in voltage-mode control, the current ramp from the magnetic elements is used for the error voltage-to-PWM conversion process. The oscillator now only serves to fix the frequency of operation. A representative current-mode control IC is shown in Figure 5.11.

Although there are five basic types of current-mode control, the popular method used in modern control ICs is called "turn-on with clock." This simply means that the oscillator turns on the power switch and it is up to the loop to turn it back off. As one can see in Figure 5.11, the level of the error voltage dictates the maximum level of peak current that will be allowed to enter the transformer and/or the output $L-C$ filter. If the input line changes, it is immediately sensed in the change in peak current within the magnetics and is held constant. If the load changes, the voltage error amplifier allows higher peak currents to enter the output filter(s). Also, the level of the valley current (I_{min}) changes. This is immediately compensated for by the current feedback loop. So the internal states of the magnetic portions of the supply are sensed and quickly compensated for.

One major problem regarding the turn-on with clock method of control is that for duty cycles greater than 50 percent the system becomes unstable. This can be corrected by what is called "slope compensation." This requires two other signals to be summed into the feedback current

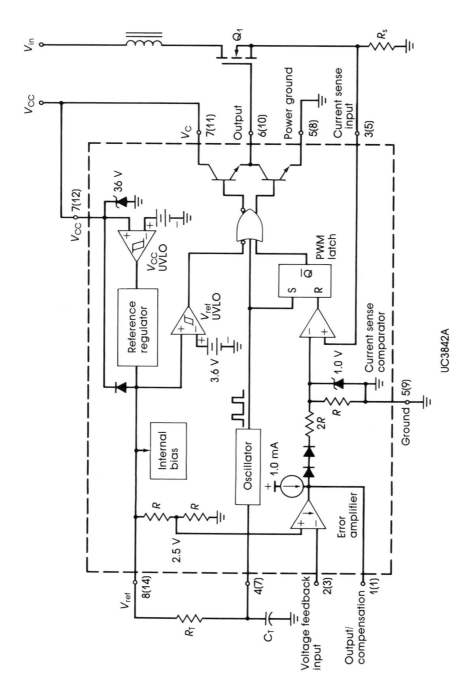

Figure 5.11

Representative current-mode control IC. Waveforms appear on the facing page (courtesy of Motorola Semiconductor, Inc.).

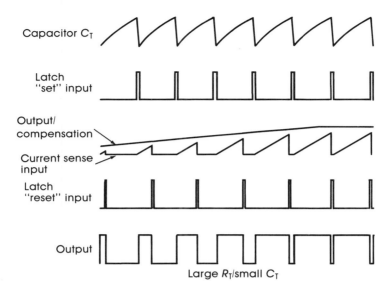

Large R_T/small C_T

ramp. The circuitry needed for slope compensation is not provided within the current-mode controller ICs, so it must be added externally. Most of the problem arises in continuous-mode flyback converters where a right-half-plane zero emerges when the continuous mode of operation is entered. Most designers prefer simply to limit the range of operation to less than 50 percent duty cycle within flyback converters. Overcurrent protection is also more tricky with the current-mode controller. Usually an overcurrent amplifier is not provided. Instead, basic current-mode control offers a constant–power type of protection (see Section 8.2).

Some of the popular current-mode control ICs are listed below.

Single-Ended Control	Double-Ended Control
UC3842/43/45	UC3825
MC34129	
MC34065	

5.4.3 Quasi-Resonant-Mode Control

Quasi-resonant switching power supplies are an emerging technology that preshapes the power switch conduction waveforms into sinusoids. This guarantees that during the switching transitions the product of volt-

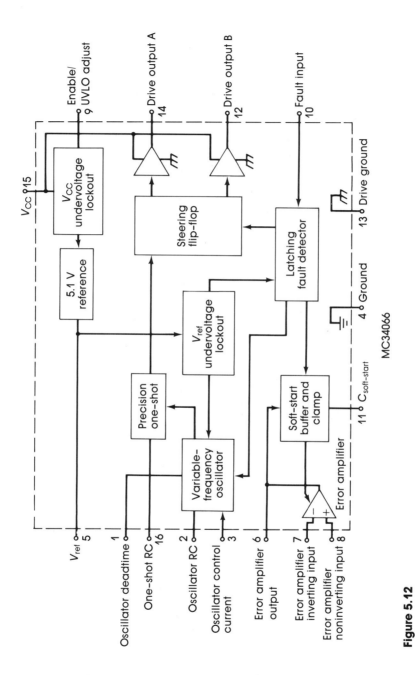

Figure 5.12

Simplified block diagram of a representative zero-current switching quasi-resonant control IC (courtesy of Motorola Semiconductor, Inc.).

age and current is zero or in other words there are zero switching losses within the semiconductors. These types of converters use one of two methods of control: (1) fixed on-time and variable off-time for zero current switching or (2) fixed off-time and variable on-time for zero voltage switching (see Chapter 11). Control is accomplished by varying the number of resonant conduction cycles per second to the output load.

Control ICs are just emerging onto the market to integrate the functions needed for these types of supplies. A representative quasi-resonant controller IC can be found in Figure 5.12. Some of the ones currently being offered are

MC34066   ZCS
LD405   ZCS
UC3860   ZCS

There will, no doubt, be more very shortly.

6

The Magnetic Components within a Switching Power Supply

The magnetic elements within a switching power supply play a cornerstone role in the operation of the supply—so much so that they are the first components to be designed, followed by the accompanying power circuits and finally by the low-level control circuitry. The typical design propagates from the magnetic elements. So a good switching power supply design first starts with a good magnetic component design and a good understanding of the basic magnetic principles.

Unfortunately, the design of any magnetic element is viewed by many engineers as black magic, and there are some valid reasons for this thinking. First, in the modern engineering community, the need for specialization has carried many engineers through a path of education where magnetic principles are only overviewed. So these engineers, who may know the *B–H* curve on sight, lack the intuitive foundation necessary to approach the design of the magnetic elements in a thorough manner. They may even be frightened off by the sight of the strange-looking equations that are associated with the magnetic phenomenon. Second, the engineer does not have good coherent information from the vendors in the field of switching power supplies. The core manufacturers present their material in terms of magnetic parameters with little or no reference to the terminal voltages or currents that can be directly viewed by the engineer. The semiconductor manufacturers do not emphasize the design of the magnetic elements, hoping that the engineer can find other design material that can provide the information. As a result, the engineer is left to sift through the readily available information, which contains large gaps in content. Hopefully this book will help bridge that gap.

6.1 Basic Magnetism and Ferromagnetism

Magnetic fields are the invisible companions to every electronics design. Since the typical workbench does not have the equipment necessary to view their presence, they often go ignored until it is time to test the product for RFI. Whenever there is a current flowing through a wire there is an associated electric field that is radially emitted normal to the current flow and a magnetic field that flows in a plane perpendicular to the current flow around the wire (see Fig. 6.1). The directions of the electric and magnetic fields are given by the "right-hand rule."

When the wire is made into a spiral coil where the adjacent section of wire is carrying current in the same direction, the magnetic fields sum together around the coil and form a combined magnetic field which flows around the coil as shown in Figure 6.2. The lines drawn to represent the magnetic field in this figure are called *lines of flux*, which is a convenient method of showing the direction of flow and density of the magnetic field around a magnetic path. Since the magnetic flux must flow around the wires carrying the current, the lines of flux become more concentrated inside the coil, where there is less area available. So this concept of the degree of concentration of lines of flux, called the *flux density* and represented by the symbol B, is expressed in units of webers per square meter [meter-kilogram-second (MKS) system] or "gauss." This can be roughly viewed as the magnetic equivalent to current density in the electrical domain. The force that creates the flow of lines of flux, called *magnetic field strength* and represented by the symbol H, is expressed in units of oersteds; H is a magnetic field strength gradient (or change in field strength) along a magnetic path roughly equivalent to the voltage drop around an electric circuit loop. It is measured in ampere-turns per meter (MKS).

Now what happens when one winds a coil of wire around a toroidal piece of magnetic material such as iron or nickel. When a magnetic field is created within the metal by passing current through the coil, some of the magnetic domains, which are small volumes formed by adjacent atoms being aligned in the same direction, will realign their dominant magnetic orientation in the direction of the magnetic field. If the magnetic field applied to the material were high enough, a point eventually would be reached where all the magnetic domains would be aligned with the magnetic field and the material would be said to be "saturated." When the field is removed, some of the magnetic domains retain their

Figure 6.1

The electric and magnetic fields surrounding a wire with a DC current flowing.

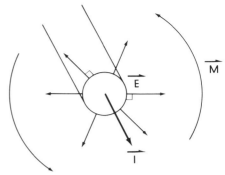

Figure 6.2

The magnetic field created by a coil.

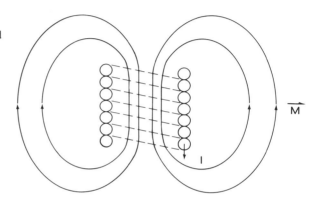

orientation from when the field was applied. This phenomenon, called *residual magnetism,* is used to create permanent magnets. Since the magnetic domains within the material are randomly oriented initially, the domains oriented closest to the direction of the magnetic field take relatively little work to reorient. The domains with higher degrees of orientation require more work in terms of a higher magnetic field strength to reorient them. This degree of work required to reorient the magnetic domains, called the *permeability* of the material, is different for every material and its alloys.

One can actually measure these parameters on the bench by using the setup shown in Figure 6.3. The resultant curve that is seen on the screen of the X–Y oscilloscope is the familiar B–H curve or hysteresis loop. From this curve one can determine the relative values of important magnetic parameters for the core material under test. To obtain accurate

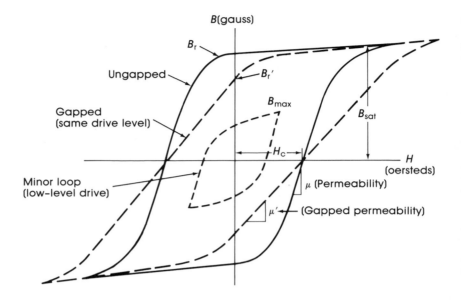

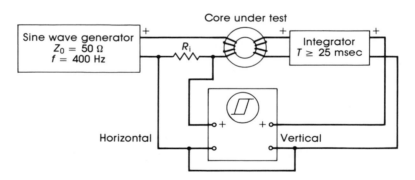

Figure 6.3

Representative *B–H* curve (*top*); method of viewing and measuring the *B–H* characteristics (*bottom*).

measurements of the parameters, an extensive effort would have to be undertaken to calibrate all the elements used in the test setup. The vertical scale is the flux density (B) and the horizontal scale is the magnetic field strength (H). The saturation flux density (B_{sat}) is the value of B at the top and bottom of the curve when the core is driven hard enough to flatten the *B–H* curve. The residual flux density is also measured while the core is overdriven and is the value of B at the y axis intercepts ($H = 0$). The slope of the sides of the *B–H* curve is the permeability

(μ) of the material. The degree of "squareness" of the material is given as B_r/B_{sat}. When the core is driven at a lower level than that required to enter saturation, the resultant curve is called a "minor loop", which shows that the permeability near the x axis and squareness do not change with drive level for a particular material. If an airgap were to be introduced within the magnetic circuit, the magnetic properties of the air would begin to influence the $B–H$ characteristic of the material. The permeability of the material drops significantly in proportion to the length of the airgap.

The quantitative values of the magnetic parameters can be determined by using the associated equations. The magnetic field strength (H) is given by

$$H = \frac{4\pi NI}{l} \tag{6.1}$$

where N is the number of turns in the winding, I the instantaneous current flowing through the wire, and l the mean magnetic length of the core. As one can see, the value of H is directly proportional to the winding current and the number of turns in the winding and is inversely proportional to the magnetic length over which the flux must travel. This also indicates that the value of H is independent of the material used within the core. The relationship between B and H is

$$B = \mu H \tag{6.2}$$

The instantaneous value of the flux density B is a little more complicated to determine, and its equation can be misused. The peak value of B in a bidirectionally driven choke (i.e., the flux is driven in both the $+$ and $-$ directions) is given by Faraday's law (rearranged):

$$B_{max} = \frac{E \cdot 10^8}{k \cdot N \cdot A_c \cdot f} \tag{6.3}$$

where B_{max} is the peak change in B, E the voltage applied to the drive winding, k a constant for the driving voltage waveform ($k = 4$ for rectangular waves; $k = 4.4$ for sine waves), N the number of turns in the driving winding, A_c the effective cross-sectional area of the core, and f the excitation frequency of the voltage.

This equation cannot be used in unidirectional applications because of the residual flux density remaining in the core prior to the beginning of the unipolar voltage waveform. This memory effect of the preceding cycles makes the value of B_{max} ride on top of a DC value of B that has been stored by the core. An equation that gives a reasonable approximation of peak excursion in B in a unipolar application such as DC chokes or flyback transformers is given by

$$B_{\text{max}}(\text{DC}) \cong \frac{L(I_{\text{pk}} - I_{\text{min}})10^8}{kA_{\text{e}} \cdot f \cdot N} + \frac{\mu I_{\text{min}}N}{l} \tag{6.4}$$

where L is the drive winding's inductance (in henries), I_{pk} the highest measured current during the cycle, I_{min} the lowest measured current during the cycle, f the frequency, A_{e} the core cross-sectional area, and l the mean magnetic length of the core. This equation gives a more accurate behavior of B in relation to the current and voltage waveforms viewed on the terminals of the choke or flyback transformer. It can also be used to plot the minor loop of the B–H curve at a given instant in time.

Now let's consider the different applications of magnetic elements as they are used in switching power supplies in relationship to the B–H curve. The transformers used in transformer-isolated forward-mode converters, such as the push–pull and half and full-bridge regulators, drive the flux in a bipolar fashion. That is, the flux is driven in both the positive and negative directions as shown for curve A in Figure 6.4. The minor loop curve is symmetrical about the origin if the winding is driven identically in both polarities. The curve may become asymmetrical about the origin if the winding is driven longer or at a higher terminal voltage in one direction. This can be caused by any differences in the turn-on or turn-off or in the saturation voltage characteristics of the power switches. Minor loop B represents a discontinuous-mode flyback transformer where the secondaries empty all the energy available in the flux and only the residual magnetism remains. This residual point becomes the initial flux in the core at the start of the next cycle. The inductor in an L–C filter, and the continuous-mode flyback transformer operate on a minor loop similar to curve C. Actually, the continuous-mode flyback transformer and the forward-mode filter choke operate identically from a magnetic standpoint except that the flyback transformer is a choke with a secondary. In this case, the excess flux is never emptied from the core. So at the start of the next power switch on-time, the flux level in the core is the residual flux plus the remaining excess flux from the last cycle. Since H is proportional to current, it is no wonder that the current at the start of a power switch conduction cycle in this mode of operation jumps up to an initial current (I_{min}) and then begins the familiar current ramp associated with the inductance.

For unipolar applications such as the flyback transformer, it is usually

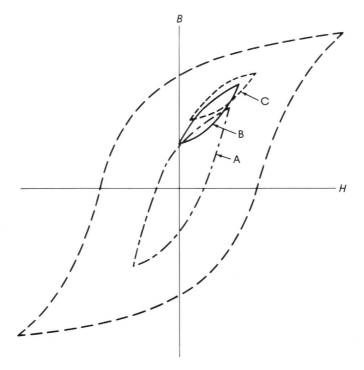

Figure 6.4

Minor loop operation of the magnetic components within a switching power supply:
(A) bipolar forward-mode transformer; (B) discontinuous flyback transformer;
(C) continuous flyback transformer and forward-mode filter choke.

desired to place an airgap in the core. This "stretches out" the $B–H$ curve along the H axis, thus making it more difficult to enter a state of saturation. The drawback is that it also requires more turns to attain the required inductance. Fortunately, inductance is proportional to the square of the turns, so the number of turns that need to be added is a smaller proportion than the increase in the amount of current required to enter saturation. The size of inductor may grow, nonetheless.

An important factor that often is ignored is the amount of loss the supply exhibits within the various magnetic cores within the switching power supply. One of the major losses is called the *hysteresis loss*. This is the energy lost in the work required to reorient the magnetic domains within the core material; it can be visualized in Figure 6.5.

The enclosed area swept out by the $B–H$ curve during one complete operating cycle of the power supply is the hysteresis loss exhibited by

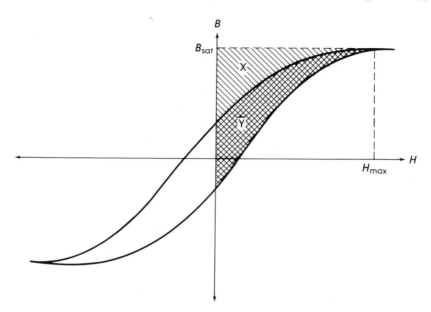

Figure 6.5

The hysteresis loss for one-half cycle. Areas $X + Y$: total energy input into the core during $\frac{1}{2}$ drive cycle; area X: energy released from the core to the output(s); area Y: hysteresis loss during $\frac{1}{2}$ cycle.

the core during that cycle. For bipolar applications it can be described in magnetic terms by

$$P_h \cong k_h \cdot v \cdot f \cdot B_{max}^2 \qquad (6.5)$$

where k, is a constant for the magnetic material (permalloy $= 0.0001$; ferrites are higher), v the volume of the core, f the operating frequency of the power supply, and B_{max} the maximum excursion of the flux density during each cycle. For unipolar applications such as filter inductors and flyback transformers, one substitutes $B_{max} - B_{init}$ for B_{max}. Obviously, this equation is not convenient for the designer to use since it involves parameters that are not readily available from the core manufacturers. The core manufacturers have conveniently provided this information in a more usable manner. What this equation does indicate, though, is that the hysteresis core losses have a very high dependence on how high the designer chooses the B_{max}. The more the designer shrinks the size of the core by operating at higher levels of flux density, the more dramatic the rise in the hysteresis losses. Also, the hysteresis loss increases in direct proportion to the frequency. In order to minimize this loss and still maintain a small size, the designer should select a core material that has the narrowest B–H characteristic. Unfortunately, the better high-frequency materials cost more, so there is a cost-versus-loss-versus-frequency trade-off. The following ferrite materials are commonly used by switching power supply designers and have met their needs.

	Material	
Manufacturer	< 50 kHz	< 1 MHz
Magnetics, Inc.	F, T, P	K, N
TDK	H7C	
Ferroxcube	3C8	3C85

The second major loss within the core is the eddy-current loss. This loss is caused by the flow of circulating magnetic currents within the material caused by rapid transitions in the magnetic flux density. This can be seen in the equation

$$P_e \cong k_e \cdot v \cdot f^2 \cdot B_{max}^2 \qquad (6.6)$$

where k_e is the eddy-current constant for the material, f the operating frequency of the supply, v the volume of the core, and B_{max} the maximum excursion in flux density during operation.

As one can see, eddy currents are proportional to the square of both B_{max} and frequency. So operating the supply at a high frequency and at high flux densities imposes severe penalties the designer in terms of much higher eddy-current losses. The amount of eddy currents depend somewhat on core shape and cross-sectional area. The traditional method of reducing the eddy current losses for a particular material has been to use laminated-type cores. This breaks up the cross-sectional area into much smaller areas, which then discourages the flow of eddy currents within the core. For solid ferrites, though, it is desired to select a material with a high reluctance (magnetic resistance). This also reduces eddy currents. The core manufacturers have also provided empirically measured loss data and presented them in easier-to-use charts so that the designer does not have to pursue a rigorous mathematical analysis of the physical situation. Once again, the ferrite materials shown in the preceding tabular list show the optimum materials used by designers within the field.

To help put the preceding discussion in perspective, in a typical power supply(30 to 50 kHz), the core losses account for only about 2 to 3 percent loss in efficiency for the power supply. This figure obviously can go higher if the designer "pushes" up the B_{max} and the frequency in order to reduce the size of magnetic components and goes beyond that which is considered reasonable.

6.2 The Forward-Mode Transformer

The transformer in forward-mode converters plays several critical roles within the forward-mode regulator, which make switching regulators more flexible and cost-effective than nonisolated and linear regulators. For example, it (1) provides DC isolation between the input power system and the loads, and (2) allows the designer to add outputs to the converter with very little added cost over the single-output version. The forward-mode transformer, though, can determine the ultimate reliability of the entire system, and therefore its design should be approached with great discretion.

The forward-mode transformer has two or more windings on the core. The winding that inputs the power is called the *primary,* and any winding that removes power is called a *secondary.* Within forward-mode transformers, both the primary and the secondary windings conduct current simultaneously. When a voltage is applied across the primary winding, the flux within the core material begins to change as

$$\frac{d\phi}{dt} = \frac{V_{pri}}{N_{pri}} \tag{6.7}$$

The secondary winding simultaneously begins to exhibit a voltage across its winding in response to variation of the flux level within the core. This results in a current flow through the secondary and through its load. The secondary then creates an opposing flux field that almost nulls out the level of flux caused by the primary. Any difference in their absolute flux induction levels is attributed to the work needed to magnetize the core material in its *B-H* characteristic and any leakage inductance of the windings themselves. Several important observations emerge at this point. The first is that since the windings are linked by the same flux, the winding voltage relationship can be shown from Equation (6.7) to be

$$\frac{V_{pri}}{V_{sec}} = \frac{N_{pri}}{N_{sec}} \tag{6.8}$$

So the transformer can be used to step up or step down the input voltage waveform to create the appropriate voltage waveforms for the proper operation of the output filter circuit. There can also be more than one secondary winding, each related to the primary by the relationship given in Equation (6.8). Also,

$$\frac{I_{\text{pri}}}{I_{\text{sec}}} = \frac{N_{\text{sec}}}{N_{\text{pri}}} \tag{6.9}$$

But what if the current through the secondary winding is limited by the load impedance? If we substitute $V_{\text{sec}} = I_{\text{sec}}(Z_{\text{load}})$ into Equation (6.8), we have

$$Z_{\text{pri}} = Z_{\text{sec}}\left(\frac{N_{\text{pri}}}{N_{\text{sec}}}\right)^2 \tag{6.10}$$

What this indicates is that the impedance of the primary winding no longer resembles that of a simple inductor but is actually dominated by the impedance of the load on the secondary. So the inductance of the windings themselves is a very minor consideration in the design of a forward-mode transformer. This can readily be seen within any transformer-isolated, PWM, forward-mode switching regulator. The current waveform of the primary winding during the power switch's conduction period is actually the current waveform of the output filter section reflected back to the primary winding.

The first step in the design of the transformer is to select the core material that will be suitable for the application. The primary concern during this step is to select a material that exhibits low core losses at the desired frequency of operation. A ferrite alloy is the dominant choice of many designers for this application. A choice from the preceding tabular list would be a safe starting point. Once the material has been chosen, the designer knows the saturation flux density of the material and can determine the desired maximum peak operating flux density.

Next the designer should select the core geometry that would serve the needs of the application. Pot cores offer good magnetic shielding since the core surrounds the windings, but this can cause the windings to operate at a higher temperature than if they were more open to the air. Variations of the pot core geometry allow more airflow past the windings, thus allowing them to operate at a cooler temperature. If cost is an issue, then an E-I or E-C ferrite core can be used. For higher power levels (>300 to 500 W), E-I or E-C laminations can be used to build a core.

Now that the core geometry has been selected, the designer can determine the optimal core size for the particular design. At this point, the designer should determine what market the power supply will be sold into. If there are safety requirements placed on the equipment, most of the burden falls on the power supply. These safety specifications affect the way the transformer is assembled and the area needed by the windings. Inevitably, the effect is to increase the size of the transformer since insulating tape must be used during the winding of the transformer. So the designer should take this added volume of the tape insulation into account. The first step is to choose the average current density (I_D) the wires must carry. Typical ranges fall between the values of 400 and 1000 circular mils per ampere. This influences how much heating will occur in the windings. The worst-case average current in the primary winding will be

$$I_{in}(av) = \frac{P_{out}/E_{ff}(est)}{V_{in}(min)} \tag{6.11}$$

Referring to the wire data in Table 6.1, the cross-sectional area of the wire used in the primary is determined. Next the designer uses an equation developed as a guide by the core manufacturers:

$$W_a A_c = \frac{0.68 \cdot P_{out} \cdot I_D \cdot 10^3}{B_{max} \cdot f} \qquad (B_{max} \cong \tfrac{1}{2} B_{sat}) \tag{6.12}$$

This yields a number that is valid for a basic two-winding transformer. If multiple output windings are required, then increase this number by 30 to 40 percent. If approval from Underwriters Laboratory (UL) or the German Institute of Electrical Engineers (VDE) is needed for the power supply, then add another 20 percent to the result. At this point, a core size can be selected by finding a core that has a $W_a A_c$ equal to or larger than the $W_a A_c$ calculated above.

Once the particular size core has been selected, the windings themselves can be determined. The turns needed by the primary can be found by

$$N_{pri} = \frac{V_{in}(nom)10^8}{4f \cdot B_{max} A_c} \tag{6.13}$$

This number now serves as the basis for all the other windings on the transformer.

In determining the number of turns required by the highest power secondary, a few items must now be considered. The forward voltage drop of the rectifiers cannot be ignored, and the maximum allowed pulsewidth of the control loop should be included. This can be done by applying the following equation:

$$N_{sec} \geq \frac{1.1(V_{out} + V_f)}{N_{pri}(V_{in}(min) - V_{sat})D_{max}} \tag{6.14}$$

Table 6.1

Wire and Winding Data: Sizes, Areas,[a] Resistance, and Current Capacities[b] for Synthetic Film Insulated Wire[c, d]

Wire Size (AWG)	Wire Area (max) (circular mils)			Ω/1000 ft	Current (mA) Capacity
	Heavy	Triple	Quad		
8	18,010	18,360	18,960	0.6281	16,510
9	14,350	14,670	15,200	0.7925	13,090
10	11,470	11,750	12,230	0.9987	10,380
11	9,158	9,390	9,821	1.261	8,226
12	7,310	7,517	7,885	1.588	6,529
13	5,852	6,022	6,336	2.001	5,184
14	4,679	4,830	5,112	2.524	4,109
15	3,758	3,894	4,147	3.181	3,260
16	3,003	3,114	3,329	4.020	2,581
17	2,421	2,520	2,704	5.054	2,052
18	1,936	2,025	2,190	6.386	1,624
19	1,560	1,632	1,781	8.046	1,289
20	1,246	1,310	1,436	10.13	1,024
21	1,005	1,063	1,170	12.77	812.3
22	807	853	949	16.20	640.1
23	650	692	778	20.30	510.8
24	524	562	635	25.67	404.0
25	424	458	520	32.37	320.4
26	342	369	424	41.02	252.8
27	272	296	342	51.44	201.6
28	219	240	276	65.31	158.8
29	180	199	231	81.21	127.7
30	144	161	188	103.7	100.0
31	117	132	154	130.9	79.21
32	96.0	110	128	162.0	64.00
33	77.4	90.2	104	205.7	50.41
34	60.8	70.6	82.8	261.3	39.69
35	49.0	57.8	67.2	330.7	31.36
36	39.7	47.6	54.8	414.8	25.00
37	32.5	38.4	44.9	512.1	20.25
38	26.0	31.4	36.0	648.2	16.00
39	20.2	25.0	28.1	846.6	12.25
40	16.0	19.4	22.1	1,079.6	9.61
41	13.0	16.0		1,323.0	7.85
42	10.2	13.0		1,659.0	6.25
43	8.4	10.2		2,143.0	4.84
44	7.3	9.0		2,593.0	4.00

[a] Areas are for maximum wire area plus maximum insulation buildup.

[b] Based on 1000 circular mils per ampere current capacity will vary according to the geometry of the unit and may range from 750 to 1200 circular mils per ampere.

[c] Includes Formvar and Poly-Thermaleze types.

[d] Courtesy of Magnetics, Inc.

where V_{sat} is the saturation voltage drop of the power switch(es), V_f is the rated forward voltage drop of the selected rectifier, and D_{max} the maximum on-duty cycle of the supply. This yields a reasonable approximation for the minimum voltage needed to develop the specified output voltage. If it is desired that the power supply fall out of regulation further below the minimum input voltage, then increase that number by the percentage of margin that is needed. Also, always round the number of turns up to the next convenient fractional turn.

Any additional secondaries should now be referenced to this highest-power secondary since it has now fixed the volts per turn rating of the secondary side of the transformer. The turns needed for the other windings are determined by

$$N_{s2} = \frac{(V_{out2} + V_f)N_{s1}}{(V_{out1} + V_f)} \qquad (6.15)$$

Now at this point, any rounding of the results for the number of turns for the additional outputs will result in an error of the second output voltage. So care should be taken during this step. The designer can adjust the number of turns on the main output, if there is enough margin, to minimize this transformer output voltage error.

Having determined the number of turns of each output needed to produce the desired output voltages, the designer must now consider the physical arrangement of the secondaries that would meet the specified needs. Three general arrangements can be considered, as shown in Figure 6.6. Each of these arrangements affects in a small way the efficiency of the supply and the volume of wire that will be used within the transformer and the transformer cost. Figure 6.6(A) is the most common form of secondary arrangement. It has the advantage of having only one diode loss during each power conduction period. The windings themselves are operated in a half-wave fashion and hence have to carry only half the average output load current, so the wire gauge can be smaller. But each half-wave section of the winding has the full number of turns to develop the output voltage. The non-center-tapped secondary arrangement shown in Figure 6.6(B) has half the turns on the core when compared to the center-tapped arrangement, but it now has two diode drops during the power conduction cycle and the wire gauge is larger. So from an efficiency standpoint it exhibits slightly more diode conduction and switching losses than the center-tapped secondary and is higher in component cost but less expensive in labor costs during the manufacturing of the transformer. The last arrangement is completely isolated

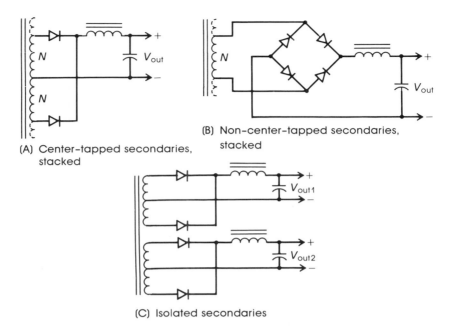

(A) Center-tapped secondaries, stacked

(B) Non-center-tapped secondaries, stacked

(C) Isolated secondaries

Figure 6.6
Arrangements of secondary windings.

outputs [Fig. 6.6(C)]. Unlike the former arrangements, which can share the turns of the lower voltage windings to develop the higher voltage outputs, each winding of the isolated version has its own windings. This arrangement can have either center-tapped or full-wave windings but utilizes an enormous amount of window area within the core. Applications that require this arrangement often require a one-point grounding scheme external to the power supply in order to reduce noise coupling between circuits. Obviously, the choice of which arrangement to use can influence both the final size of the transformer and the efficiency of the finished switching power supply.

At this point in the "paper design" of the transformer, it is a good idea to check the window fillage of the windings to verify that the selected size core will be large enough. First the designer must determine the wire gauges of the secondaries. If windings share sections of the windings, such as a +12-V winding including a +5-V winding, then the currents of both the +12-V and +5-V windings must be summed in order to determine the current flowing through the +5-V winding, and so on. To determine the total area taken up by the physical windings one simply multiplies the number of turns of a gauge wire times the area of the wire and then sums all these subproducts.

$$W_a = \sum_1^i N_i \cdot A_w(i) \qquad (6.16)$$

If the wire area is greater than 85 percent for a transformer that does not have to meet dielectric requirements, or is greater than 75 percent for one that does have to meet safety requirements, then the core may be too small. If this is the case, then the designer may have to select the next larger core size and recalculate the required turns for each winding.

This completes the "paper design" of the transformer. Additional calculations can be performed to estimate the core and winding losses in its final operation. But now the physical winding of the transformer should be considered. The major influences on the designer during this stage are electrical performance and safety considerations. Usually these two factors are at odds with one another. Factors that enhance electrical performance are tight magnetic coupling of the windings to the core, which reduces leakage inductance; tight magnetic coupling between all the windings, which aids in coupling high-frequency edges between windings; and a low voltage difference between adjacent turns, which minimizes the effects of self-interwinding capacitance. In essence, the windings want to be in very close proximity to one another and to the core. Unfortunately, the safety regulations require the primary and secondary windings to be spaced apart by insulation. This greatly degrades the interwinding coupling and leakage inductance. So the physical design of a transformer always becomes a trade-off between the safety regulations and electrical performance. Some practices that enhance performance within the physical constraints of the safety regulations are:

1. Use bifilar windings that do not have to meet dielectric tests between themselves. This means that the wires should be twisted together prior to their winding on the core. These are usually the secondary windings. This enhances the cross-regulation of the outputs.
2. Interleave the primary with the secondaries. This can be seen in Chapter 7, Figure 7.2. This enhances the cross-regulation and interwinding coupling from primary to secondary.

Figure 6.7 shows how one would wind a VDE-approved transformer. As one can see, not only do the windings have to be separated by the insulating tape, but the windings themselves must be spaced away from the ends about 2 mm to meet the creepage requirement of VDE. Also, any leads that pass through other windings and are not of the same

Figure 6.7
Designing the physical
transformer to meet safety
requirements.

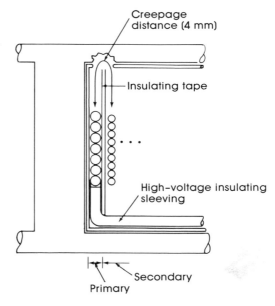

Creepage
distance (4 mm)

Insulating tape

High-voltage insulating
sleeving

Secondary
Primary

dielectric rating must be covered with a high-voltage insulating sleeving that meets or exceeds the dielectric rating. All of these practices waste valuable wire winding area, which makes the transformer physically larger. For those power supplies that have all voltages less than 42.5 V and all currents less than 8 A, these safety considerations need not apply. All the windings could be bifilar wound as much as possible to enhance the electrical performance.

As one can appreciate, the design of a forward-mode transformer can be an iterative process. Keep in mind that the design equations produce parameters that are best estimates and should not be interpreted as precise values.

6.3 The Flyback Transformer

The flyback transformer is treated quite differently from the forward-mode transformer in the design process. It more closely resembles a forward-mode choke, and, in fact, the flyback's original name was a "swinging choke" regulator. If one were to look at the time function of the flux within the core, one couldn't tell whether the core belonged to a forward-mode filter choke or a flyback transformer. This can be illustrated by referring to Figure 4.5 and superimposing the primary and

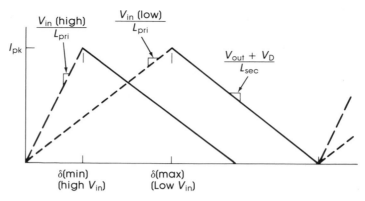

Figure 6.8

Superimposed primary and secondary currents in a flyback transformer.

secondary current waveforms as shown in Figure 6.8. As one might notice, they form the familiar sawtooth current waveform seen in a forward-mode filter choke.

The flyback transformer is constructed from a core with a primary winding and one or more secondary windings. For basic description purposes we will analyze only the single-output transformer and describe the method of adding additional outputs later in the section. During the design process, the flyback transformer is treated as two separate inductors that just happen to share the same core. This can be easily justified since the transformer operates with neither of the windings being active at the same time. The primary serves to store energy from the input power source within the flux of the core. When the primary has completed this process, the secondary begins to conduct and removes this stored energy, placing it in the output filter–storage capacitor. Because of this time multiplexing for the core, there is a high degree of flexibility in determining the primary–secondary turns ratio. A 1:1 ratio transformer can create 10 V as well as it can create 5 V from the same input voltage. Likewise, a 1:2 transformer can create the same voltages as a 1:1 turns ratio transformer does. This might give rise to some confusion within the mind of the designer, but if the designer appreciates some of the other design considerations that are related to the turns ratio, then the confusion can be avoided.

The first and most important factor in the design of a flyback transformer is that the primary be capable of storing enough energy within the core in the permitted time period to maintain the necessary power required by the load. Most designers allow 50 percent of the power supply's operating period at the worst-case input and loading condition.

This occurs at the lowest specified input operating voltage and at the full-rated output load. When the power switch is turned on, the current through the primary begins to ramp up at a rate given by

$$i_{pri}(t) = \frac{1}{L_{pri}} \int_0^{T_{on}} V_{in} \, dt \qquad (6.17)$$

or

$$i_{pk} = \frac{V_{in} \cdot T_{on}}{L_{pri}} \qquad (6.18)$$

As one can see, the peak current in the primary is dependent on the input voltage and the on-time of the power switch. This is where the control circuit's influence enters the picture. The power stored in the core is proportional to the square of the peak current, so as the input voltage varies, the on-time of the power switch must also change in order to maintain the desired primary peak current. So two fundamental parameters emerge as being important: (1) the stored energy as given by

$$W(t) = \tfrac{1}{2} L_{pri}(i_{pk} - i_{min})^2 \qquad (6.19)$$

and (2) the maximum time that is allowed to store that power $[T_{on}(\max)]$. The steady-state power is given as the energy stored during each on-time multiplied by the frequency (to normalize it to one second):

$$P_{pri} = \tfrac{1}{2} L_{pri}(i_{pk})^2 \cdot f \qquad (6.20)$$

Before we can create a nice "turn-the-crank" equation, one additional factor must be considered. The flyback transformer can operate in two different modes: (1) the discontinuous mode, where all the energy within the core is emptied by the secondary and (2) the continuous mode, where some of the stored energy remains in the core. The difference can be seen in Figure 4.6. For the discontinuous mode, the primary current starts from zero [or the minimum current in Eq. (6.19) equals zero]. For the continuous mode the current ramp rides atop a pedestal pulse, which causes the minimum current to begin at a nonzero value. The vast majority of flyback converters are designed to operate in the discontinuous mode for two reasons: (1) the transformers can be made smaller since this requires a smaller primary and secondary inductance, and (2) the continuous-mode converter is more difficult to stabilize and has a smaller operating range. Most discontinuous-mode flyback converters, though, will enter the continuous mode of operation at its low input voltage condition just prior to falling out of regulation.

So knowing that we want to design a discontinuous-mode flyback converter, we can combine the two important equations [(6.18) and

(6.19)] where the minimum current equals zero, and produce the relationships that satisfy the two important parameters of time and power. First the minimum peak current necessary to store the needed power within the core is given by

$$i_{pk}(\min) = \frac{2P_{out}}{V_{in}(\min) \cdot \delta_{(max)}} \qquad (6.21)$$

From this we can derive the largest primary inductance necessary to produce that peak current needed within the allowed time period:

$$L_{pri}(\max) = \frac{V_{in}(\min) \cdot \delta_{(max)}}{I_{pk} \cdot f} \qquad (6.22)$$

This calculation is performed at the lowest specified input voltage and the full-rated load. If the secondary requires the balance of time to empty the power, the flyback converter will fall out of regulation just below this input voltage. It is recommended that a small amount of voltage margin be included in this calculation so that the designer can guarantee regulation while including the tolerances in the construction of the transformer.

At this point we are ready to select a core and begin the design of the physical transformer. It is also appropriate at this point to pass on some pointers unique to flyback transformer core selection. Since flyback transformers store relatively large amounts of energy in very short periods of time, the inductances are quite low. What this translates to is that the transformer will enter saturation in a much shorter on-time than will inductors and transformers used within forward mode converters. Flyback transformers usually are designed with very little margin with respect to time and can easily run into saturation problems at the high input voltage line with a step increase in load current. This situation causes the error amplifier to rapidly increase the pulsewidth to its maximum, and the volt–time product at the high input line may be sufficient to cause the saturation of the transformer core. The airgap "stretches" the B–H curve along the H axis, which indicates that it now requires more primary current to reach the saturation point of the core material. This can provide the needed margin in the transformer and avoid a possible power transistor or MOSFET failure during its operating life. To avoid this situation, an airgap should be added to the core. The airgap can offer some problems to the designer, who now must add turns to the transformer in order to achieve the desired inductance, which can cause the size of the core to increase. But since the inductance increases as the square of the turns, the number of turns required to reattain the inductance is less than the effects of introducing the airgap. What this means during the core selection process is that the designer should consider only cores that have airgaps in their family members. First, molybdenum–permalloy powder cores are self-gapped by the introduction of nonmagnetic particles within the magnetic material. The windings themselves also couple better to the core and to the other windings with the toroid core. Unfortunately, the cost in labor to wind the windings on the toroid can be 2 to 3 times that of a bobbin-style core. Since cost is usually a dominant factor in the design of a switching power supply, bobbin-style, ferrite cores are usually selected. These are pot cores, E-C cores, or other bobbin-wound ferrite cores that have air-gapped family members. Finally, the core material should exhibit low hysteresis losses at the high flux densities at which flyback transformers operate and at their high frequency of operation.

Core manufacturers have universally adopted a mathematical method of easing the core selection process. The method results in the product of the required wire cross-sectional area needed to couple the electrical power into the core times the required core cross-sectional area needed to support the generated magnetic energy. This parameter can be calculated for a single winding by the following expression:

$$W_a A_c = \frac{(6.33 \ L_{pri} I_{pk} \cdot W_a)10^8}{B_{max}} \tag{6.23}$$

where W_a is the area of the primary winding (in circular mils) needed to support the worst-case average current (obtained from the wire gauge chart; see Table 6.1) and B_{max} is the maximum excursion of the peak operating flux density (usually half of B_{sat} at the nomimal input voltage).

Because flyback transformers have secondary windings included, this equation needs some modification in order to provide the needed space for the added windings. Theoretically, the wire area occupied by the secondary should be the same as the primary since the power removed from the core equals the power placed in the core. In reality, though, the secondary is usually composed of more than one winding, each of a different wire size, which will inevitably lead to winding inefficiencies. So usually the secondary will occupy between 60 to 70 percent total winding area. So if Equation (6.23) is modified to accommodate the secondaries, the expression becomes

$$W_a A_c \cong \frac{(25 L_{pri} I_{pk} \cdot W_a(pri))10^8}{B_{max}} \tag{6.24}$$

The next step after selecting the core style and material is to select the mimimum core size needed for the application. This is done by selecting a core size that is just larger than the $W_a A_c$ calculated in Equation (6.24). It may also be a good idea to request samples of not only the selected core size but also the next larger core size in case the windings do not fit in the first core.

Next the estimated airgap should be calculated by

$$l_g(\text{est}) = \frac{(0.4\pi L_{\text{pri}}I_{\text{pk}})10^8}{A_c B_{\text{max}}^2} \qquad (6.25)$$

The core area A_c is calculated from the core data sheet, and B_{max} is typically chosen to be one-half the saturation flux density of the core material used. Next select the core part number within this size whose airgap comes closest to the calculated airgap from Equation (6.25). From this point on, the actual airgap from the core specification sheet should be used in all following calculations.

Now that the physical core has been determined, one can begin calculating the number of turns needed for the primary winding. This can be obtained by

$$N_{\text{pri}} = \frac{B_{\text{max}}l_g(\text{act})}{0.4\pi \cdot I_{\text{pk}}} \qquad (6.26)$$

At this point, it is always advisable to double-check the peak excursion of the flux density at the input voltage at which the supply is expected to operate most during its life. This can be done by the use of Faraday's law (rearranged):

$$B_{\text{max}} = \frac{V_{\text{in}}(\text{nom}) \cdot 10^8}{4 \cdot A_c \cdot N_{\text{pri}} \cdot f} \qquad (6.27)$$

The resulting B_{max} should be close to 50 percent of the saturation flux density. If it exceeds 65 to 70 percent, the designer is advised to increase the airgap and recalculate the needed turns. Operating with too high a B_{max} will lead to problems at the upper end of the specified input voltage and at high temperatures.

The turns required for the secondary can now be determined. Remember, the secondary must remove all the stored energy from the core prior to the beginning of the next power switch conduction period. So the minimum time period for this to happen is once again at the minimum specified input voltage line when the power switch's conduction period is at its maximum. So the number of turns can be obtained by

$$N_{\text{sec}}(\text{max}) = \frac{N_{\text{pri}} (V_{\text{out}} + V_{\text{D}})(1 - \delta_{\text{max}})}{V_{\text{in}}(\text{min}) \cdot \delta_{\text{max}}} \qquad (6.28)$$

where V_{D} is the forward voltage drop of the rectifier.

If the calculation results in a noninteger number of turns, round downward to the closest half or quarter turn, whichever is dictated by the core construction. Never have a winding of less than one turn. This will result in very poor coupling to the core and to the primary.

Addition of outputs to the power supply is an easy matter at this point and is similar to adding turns to a forward-mode transformer. The designer would use the volts per turn of the main winding and calculate the other windings. Remember that the winding voltage must include the forward voltage drop of the rectifier. The number of turns for each additional secondary can be determined by

$$N_2 = \frac{(V_{out2} + V_D)N_1}{(V_{out1} + V_D)} \tag{6.29}$$

where V_{out1} is the main output voltage of the power supply (the highest power output) and V_{out2} is the desired additional output voltage. If the resultant number of turns is a noninteger and the turns must be rounded to the closest number that the core construction allows, there will be an error in the absolute accuracy of that output voltage. If the error is unacceptable, the designer can vary the turns of the main output winding until an acceptable error on all the additional outputs is obtained.

Now at this point, the "paper" design of the transformer is complete, but this is only half the job. Because the flyback transformer stores a large amount of energy during each cycle, it has the capability to produce large, high-energy voltage spikes on its windings. A poor physical design of the transformer will create voltage spikes on the primary that can easily exceed 200 V above the input voltage plus the flyback voltage. This can cause the power transistor or MOSFET to go into avalanche and destroy the part. So the "bandaid" remedy for this is to add a snubber across the primary winding that will reduce the supply's operating efficiency by several percent. There is much that can be done to reduce this spike in the physical design of the transformer itself.

Two factors affect the amplitude of the spike. The first is the coupling of the primary winding to the core itself. Loosely wound windings will permit some of the wire's generated magnetic field to circulate only in the airspace provided by the winding volume. This resembles an additional inductor in series with the inductance of the primary. This is called *leakage inductance,* and it can store energy of its own, which can cause spiking. If the winding is particularly loose, not enough energy will be stored in the core and the load-handling capability of the entire supply will suffer. The second factor is the capacitive coupling between the primary and secondary and between secondaries. The primary-to-secondary coupling is important to transmit the rapid transitions of the primary to the secondary so that the rectifier can see the turn-on voltage faster. The time between the power switch's turn-off and the diode beginning to conduct current leaves the "charged" core completely unloaded, so very high voltage spikes are created. Bifilar winding of the primary and secondaries, if the input voltage does not exceed 42.5 V, or interleaving the windings for higher input voltages is recommended. Bifilar winding of the secondaries is also recommended in order to enhance the cross-regulation of the supply. The capacitive coupling for the interleaved windings is worse so a bifilar wound, clamp winding may be recommended for the primary.

Many companies decide to have the transformers manufactured by an outside transformer manufacturer. These companies are driven by labor costs and will recommend "cheaper" methods of winding the transformer. I heartily recommend specifying the winding method as well as the desired winding inductances. It may cost more, but the beneficial effects on the reliability of the supply will be worth it.

6.4 The Forward-Mode Filter Choke

The forward-mode filter choke forms the backbone of all the forward-mode converters. It is always used as part of a choke input filter, which is composed of the series inductor and a shunt capacitor. It is easy to calculate the inductance needed by the circuit, but the core selection process can be confusing to the designer because it can require a few iterations to finally arrive at the final choice.

The inductor within a choke input filter is operated in the continuous mode (an expression borrowed from the flyback transformer); that is, the core must never be emptied of its energy. Unlike the flyback converter, where all the output energy is stored in the output capacitor, the task of output load energy storage is shared between the inductor and the capacitor. The typical choke is designed to have 50 percent more energy than what is required by the load during one cycle of operation. The amount of energy that enters and leaves the inductor's core during each cycle is given by

$$\Delta E = \tfrac{1}{2}L(i_{pk} - i_{min})^2 \tag{6.30}$$

and the amount of energy remaining in the core is given by

$$E_{resid} = \tfrac{1}{2}Li_{min}^2 \tag{6.31}$$

The amount of energy that resides in the output filter capacitor is

$$E_{cap} = \tfrac{1}{2}C_0 V_{out}^2 \tag{6.32}$$

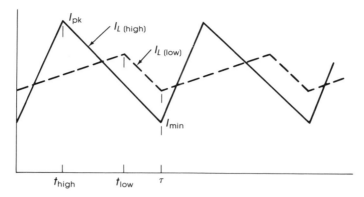

Figure 6.9
Current within a forward-mode inductor.

In a typical application the average energies stored within the inductor and the capacitor are about equal. This may not be suitable for certain applications because of the nature of the load. The energy within the inductor cannot be instantaneously removed by the load, since it resembles a current source. So for loads that exhibit step changes in their current demand, the supply's output would exhibit a long and pronounced dynamic load response. This is where the output filter circuit cannot provide the immediate current demand of the load until the control loop can compensate for the increase in load current. The capacitor can provide this instantaneous current to the load, so the designer should either increase the value of the capacitor above what is indicated by the design equations and/or slightly decrease the value of the inductor below what is calculated. This will make the output resemble more of a voltage source and allow the control loop to respond to the transient changes in the load more quickly.

Before calculating the minimum L value of the filter inductor for a constant-current type load such as logic or microcomputers, the designer must first know the maximum input voltage, the step-up or step-down transformer's turns ratio, and the minimum expected load current for this output (see also Fig. 6.9). The worst-case demand for the inductor occurs at the highest input voltage where the on-time duty cycle of the power switch is at its minimum. If one is not certain of the expected on-time duty cycle of the switching supply at the high input voltage, 10 percent is a reasonable guess. The minimum inductance value is then obtained by

$$L_{min} = \frac{V_{in}(max) \cdot T_{off}(max)}{1.4 \cdot I_{out}(min)} \tag{6.33}$$

The V_{in} is the actual voltage presented to the input of the inductor, which is either V_{in} minus the saturation voltage of the power switch for buck converters or the transformer's output voltage minus the voltage drop of the rectifier. The resulting inductor value will assure the designer that the inductor will not run out of energy at any point during the supply's operation. It is recommended that adjustment of the inductor's value for transient-type loads be done on a working model and the inductor's current waveform be observed during the process.

The core can now be selected. Because of the DC bias on the core a gapped ferrite core or a permalloy powder core toroid should be used. A gapped ferrite core may be used, and one would use Equation (6.23) to determine the required W_aA_c, then determine the needed gap by Equation (6.25, and finally determine the required turns by Equation (6.26). Many designers, though, use permalloy powder toroids for this application since the labor costs to place one winding on the toroid is much less than that of a transformer. They also exhibit better core coupling characteristics and lower leakage inductance. Unfortunately, there is no adequate method for determining the proper core size and permeability used by the core manufacturers. The process usually requires a few iterations before the smallest possible core size can be determined. The method used by Magnetics, Inc. (see also Fig. 6.10) is given below.

1. Calculate the product:

$$E_{av} = LI_0^2(av) \qquad\qquad (6.34)$$

   where L is the minimum inductance from Equation (6.33) and I is the average rated load current of the output.
2. Refer to Figure 6.10 and locate the value of the above product on the x axis (horizontal). Proceed up a vertical line from the x axis until it intersects the first sloped function. This is the recommended permeability of the core for this application. Proceed up to the next horizontal grid line and read the basic core number from the y axis (vertical). This is the smallest core that can sustain this needed flux density.
3. Refer to the data sheet for that core and calculate the number of turns needed to create that inductance by using the following equation:

$$N = 1000\sqrt{\frac{L_{desired}}{L_{1000}}} \qquad (L_{1000} \equiv A_L) \qquad (6.35)$$

4. Determine the wire size needed from the wire table (see Table 6.1) and multiply the number of turns by the wire area and determine the

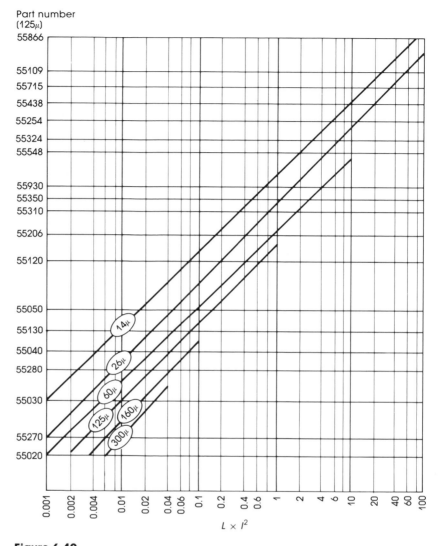

Figure 6.10
DC bias core selector chart (L = inductance with DC bias in millihenries; I = DC current in amperes). (Courtesy of Magnetics, Inc.)

percent of window fillage (i.e., how much of the hole is taken up by the wire):

$$\% \text{ window} = \frac{N \cdot A_w}{A_{wc}} (100) \qquad (6.36)$$

At this point, if the percentage of the window is greater than 40 to 50 percent, the designer should consider a larger core size. The reason for this is that the coil manufacturer needs about 50 percent of the window area to fit the winding shuttle through the window during the winding process. If the core size is too small, select the next larger core size and repeat steps 3 and 4 until a 40 to 50 percent window fillage is obtained.

For output current in excess of 5 to 10 A and frequencies greater than 30 kHz, the use of Litz wire should be considered, in order to reduce the AC resistance of the winding caused by the skin effect.

6.5 Mutually Coupled Forward-Mode Filter Inductors

Multiple-output, forward-mode converters have one major annoyance: the sheer number of inductors needed to filter each output of the supply. This can make the power supply heavy and occupy a large volume. One method to reduce the amount of core material within the supply is to employ mutually coupled filter inductors. This method basically places two output filter inductors on one core. There is an obvious space advantage in doing this. There is also a great improvement in cross regulation of those two outputs (see Chapter 7).

This method can theoretically be extended to all the outputs, but the number of turns for each winding is so critical that it cannot be economically maintained in a production environment. So the use of these windings is recommended only for outputs with equal voltages with opposing polarities (e.g., $+12$ and -12 V). This makes the turns for both windings equal and hence easy to control in production.

To design a mutually coupled inductor is as easy as it is to design a single winding inductor. The procedure is as follows.

1. Select the "complementary" outputs that are to be included in the mutually coupled inductor.
2. Calculate the minimum inductance L_{min} for the lowest current output of the two using Equation (6.33).
3. Select the core size by calculating the average energy entering the core by

$$E_{av} = L(I_{out1} + I_{out2})^2 \qquad (6.37)$$

and follow the procedure outline in Section 6.4 to determine the minimum core size.

4. Calculate the number of turns for the lowest current output by using Equation (6.35). The number of turns needed by the higher current winding is the same as the lower current winding.

5. Check the percentage of the window fillage by first determining the wire sizes needed for each winding and then by using the equation

$$\% \text{ window} = \frac{N(A_{w1} + A_{w2})}{A_{wc}} (100) \qquad (6.38)$$

where A_{wc} is the window area of the core. Once again, if the percentage of window fillage is greater than 40 to 50 percent, a larger core should be used.

This core is easy to wind physically. If one twists the wires together prior to winding them onto the core, the windings are guaranteed to be of equal turns, which is very critical. When the mutually coupled choke is placed within the circuit, the polarities of the windings must also be reversed since their voltages are opposite, and preferably their fluxes should add rather than subtract. The result is a mutually shared magnetic reservoir to which both transformer windings contribute equally. It also reflects any changes in the load on the negative winding, which seldom is directly sensed, to the positive winding, which can be directly sensed. This causes the load regulation of the negative output and the cross-regulation of both windings to improve dramatically. I would personally recommend mutually coupling as many complementary output filter chokes as possible to save space and to improve cross-regulation.

7

Cross-Regulation of the Outputs

Cross-regulation is how all the outputs of a multiple-output switching regulator respond to a change in the load on any one output. Cross-regulation is not a commonly discussed subject, but is nonetheless important to the operation of a well-designed switching power supply. The primary areas within the supply that affect how well it will exhibit cross-regulation are the magnetic components and the method by which the outputs are sensed.

The phenomenon of poor cross-regulation can be described simply. First, in most multiple output switching regulators, only one or a few of the output voltages can be sensed. It is impractical and costly to invert and/or scale each output voltage to meet the input levels of the error amplifier. What this means is that the sensed output has very tight load regulation, and the other outputs exhibit much poorer load regulation. Any changes in the load current of the sensed output are immediately sensed by the error amplifier/PWM-generating loop, thus maintaining its output voltage. The other outputs do not experience the change in load current but still have a different pulsewidth presented to them that results in a different output voltage value. Typically, when an increase in load occurs on the sensed output, the voltages of the other unsensed outputs increase, sometimes by more than 20 percent. Conversely, when an increase in load current occurs on an unsensed output, the sensed load stays solidly at its rated voltage and the unsensed output experiencing the load, exeriences a reduction in voltage. To a load, this unexpected supply voltage variation can lead to erratic operation or even failure (see Fig. 7.1).

Ideal cross-regulation cannot be practically attained short of placing a linear postregulator on each output, which compromises the supply's efficiency. What happens is that all the outputs exhibit a worsening of

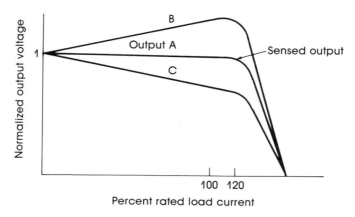

(A) Supply exhibiting poor cross regulation

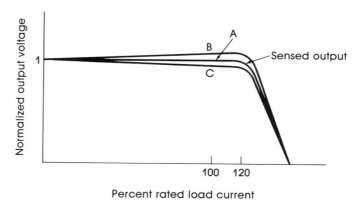

(B) Supply exhibiting good cross regulation

Figure 7.1

Graphical representation of the degree of cross-regulation in a multiple-output switching supply.

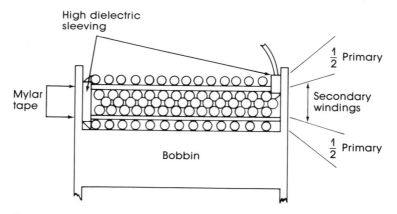

Figure 7.2

Winding interleaving for high-dielectric isolation and coupling.

load regulation, but the degree of voltage variation on the unsensed outputs is greatly improved.

The three areas inside the supply to consider during the design phase for improved cross-regulation are the transformer, the output filters, and the output-voltage-sensing network. By using one or more of the following techniques, one can drastically improve the supply's cross-regulation.

7.1 Transformer Techniques

The transformer is the heart of the switching regulator, and its design should be of primary concern. The engineer should not only compute the turns of each winding and determine into which core the windings will fit but also consider the method by which the windings are wound on the core. First, all windings that do not have to meet a high dielectric voltage isolation requirement should be bifilar-wound. This means twisting the winding wires together before they are wound on the core. This gives excellent coupling between those bifilar-wound windings and also drastically reduces the leakage inductance of the windings. This coupling is important because it transfers the rapid switching edges between the primary winding and the secondaries. This makes the rectifiers work more efficiently and hence the rectified pulsetrain has a more authentic appearance, and any changes in output load of any of the outputs is reflected back through the core and sensed by the voltage-sensing network. If a high dielectric isolation requirement is necessary (off-line regulators) between the primary and secondaries, a technique called interleaving should be used. This is where one-half of the primary winding is placed next to the core. This layer is then covered with a layer of Mylar tape (for isolation), and then the bifilar wound secondaries are placed next on the core and covered with another layer of Mylar tape. Finally, the remainder of the primary winding is placed on the outside. This sandwiches the secondaries between the primary layers, thus permitting the closest coupling possible in this situation. It is not as good as bifilar-winding all the windings, so one should expect some inductive spiking in the primary winding (see Fig. 7.2).

7.2 The Voltage-Sensing Network

This technique yields a very dramatic improvement in output cross-regulation and is very inexpensive to implement. Essentially, it is just

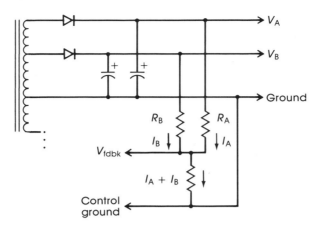

Figure 7.3
Multiple-output sensing network.

sensing two or more of the positive outputs. This is done by adding
another resistor to the resistor divider from the high current output (see
Fig. 7.3).

This technique basically creates a current-summing node at the input
of the voltage error amplifier. Most of the sensing current would be
taken from the highest-power positive output. A lesser amount is taken
from another lower-power positive output. The percentage of sensing,
which influences how tightly each output is regulated, is directly dependent on the percentage of the total current entering the summing node
from each output.

The benefit this technique offers is that it not only regulates the two
output voltages as referenced from the output ground, but it also regulates the difference between the two outputs. This difference sensing
actually allows the error amplifier to detect changes in the magnetic
parameters within the core and compensate accordingly. The net result
is that all the outputs exhibit improved cross-regulation characteristics
regardless of whether or not the outputs are sensed. Typically only positive outputs are sensed since it is much more inconvenient to invert
negative outputs by using an additional operational amplifier.

7.3 Mutually Coupled Output Filter Chokes

This technique, theoretically, is the answer to all prayers with respect to
cross-regulation improvements in forward-mode switching regulators.

Basically, the output filter chokes for two or more outputs are wound on one core. Each winding is sized to contribute an identical amount of flux within the core. This mutual reservoir of flux is able to be drawn on by each output resident on the core. One of the outputs on the core is sensed by the error amplifier. When the load of one of the unsensed outputs changes, it is reflected through the filter choke core and imme- diately detected by the error amplifier. Also, if one of the outputs goes into an overcurrent condition, all the outputs fold back in an identical fashion. This makes the supply behave the way one thinks it ought to behave (see Fig. 7.4).

There is one major problem with this technique: it is convienient only for two outputs of identical voltage magnitude (not polarity) to be fil- tered in this fashion. The complexity of the mathematics and the re- quired precision of the windings on the mutually coupled core must be so exact that it is impractical to maintain in a production environment. For each turn in error from optimum, one will lose about 1 percent in supply efficiency. As a matter of fact, the only way to optimize the windings on the core is to maximize the effeciency of the supply. This is an extremely laborious exercise, and then the question of whether a transformer vendor or department can maintain this accuracy in a high- volume environment is important.

The design procedure for a mutually coupled choke for identical volt- age magnitudes is to first determine the minimum required inductance for the lightest load and then select the core (size for about a 30 percent fillage) and calculate the turns required. Make the turns for the second

Figure 7.4
Mutually coupled output filter chokes.

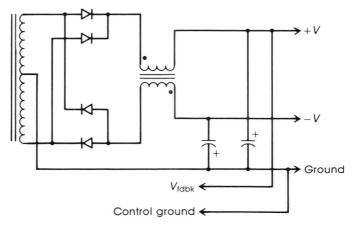

output identical. Bifilar-wind both windings on the same core, and check for the proper winding polarity during the installation process.

It is recommended that a separate core be used for each complementary pair of outputs (e.g., $+/-12$ V) and the positive outputs sensed using the multiple-output sense scheme described previously.

By using a combination of the cross-regulation improvement techniques mentioned herein, one can obtain very satisfactory performance from a multiple-output switching supply.

8

Protection

8.1 Protecting the Supply and the Load from the Input Line

Designing a switching power supply to only provide power to the load circuitry is only half the job of designing a switching power supply. The designer has not entirely fulfilled the responsibility of completing the design and the project. The design at this point addresses only the well-known, benign, steady-state operating conditions that occur during 99.99 percent of the supply's operating life. But what about that fractional percentage of conditions that was not addressed? This small percentage is what nightmares are made of. These are the adverse operating conditions and failure conditions that occur infrequently, but when they do occur, people are not likely to forget them. The much revered Murphy has a particularly appropriate saying relevant to switching power supply design: "It takes 98 percent of the time to solve 98 percent of the problem. It also takes 98 percent of the time to solve the last 2 percent." This last 2 percent in this case is the adverse conditions under which the power supply must operate. If they are not addressed by the switching power supply designer, the results usually are unexplainable field failures, which may result in a catastrophic outcome. The responsibility, rightly or wrongly, ends up with the power supply designer.

Assuming that the power supply was designed conservatively, that is, that no components are operating just below their maximum ratings, one should consider the abnormal conditions originating outside the switching power supply. These basically can be grouped as input adverse operating conditions and output (or load) adverse operation conditions, each of which may require a slight redesign of the basic power supply.

8.1.1 AC Line Input Adverse Operating Conditions

These types of adverse operating conditions indicate that the power supply designer should know something about the nature of the input power source and its distribution system. In a typical AC power distribution system as provided by the local power utility, the main adverse operating conditions are line dropouts, brownouts, surges, and transients. Each of these conditions individually can cause a failure in your supply.

AC Line Dropout

This condition usually occurs when a heavy AC motor turns on within a factory or home AC power system and the startup current surge cannot be supported by the wiring. It can also occur in an aircraft power system or in systems backed up by uninterruptable power supplies (UPSs) that use relays to switch in the backup system. The AC voltage literally disappears for a cycle or more, and then just as suddenly returns. The supply must have a large input bulk filter capacitor to allow it to continue to operate through this dropout and shut down if it continues too long. If the supply attempts a restart too early after the input power returns, the control and power switch driver supply voltages inside the supply may be at too low a level to guarantee reliable operation. The supply must also be able to restart itself as if from a "cold" turn-on condition.

Brownout Conditions

This is a condition where the input line goes below its minimum specified value for voltage. If the product were allowed to continue to operate during this condition, the outputs would likely be out of regulation (i.e., at a lower voltage). This would make the load operate below its specified voltage limits and cause erratic operation (perhaps with destructive results). Also, the control and power switch driver voltages within the supply would be operating at too low a level and would cause power switch overdissipation.

Both line dropout and brownout conditions require that the supply sense the input voltages and inhibit its operation below a certain input voltage that is above the point of destruction of the supply and/or load.

Surges and Transients

In such conditions the AC line goes above the maximum specified voltage value. These conditions can cause some components within the supply to exceed their absolute maximum voltage ratings, which may subsequently cause their failure. They are typically caused by lightning strikes on the power grid or large AC inductive loads on the system such as motors. The typical method for addressing this problem is to clamp the voltage to a safe value so that the supply does not see it. This is where a little experience goes a long way: one cannot place the clamping device directly on the input line; the energy contained in these surges are simply too large for the device handle. These devices would fail (typically short-circuit) the first time a surge enters the system. The best solution is to place a small amount of impedance in series between the line and the clamping device. The input EMI filter is a perfect choice for this. If the transorb or metal oxide varistor (MOV) is placed after the input EMI filter, some of the energy can be stored and averaged in the series inductor and safely shared with the bulk filter capacitor. The result is that the peak voltage is drastically reduced, the amount of instantaneous energy the clamping device sees is reduced, and everything stays within its rating.

8.1.2 DC Line Input Adverse Operating Conditions

The DC line has many of the same adverse operating conditions as the AC line, although the power system now may be composed of a battery and a generator (or alternator), or may be a bulk AC-to-DC power supply providing DC power to many boards. One can assume in the AC-to-DC bulk power supply system that the bulk supply provides the DC isolation from the AC line. Its adverse modes of operation are just over and under voltage conditions. The generator (alternator)–battery systems found in automobiles and aircraft constitute a more hostile environment. This system's adverse operating modes are undervoltage, overvoltage, dropout, and surges. Special care should be taken when addressing these adverse operating modes during the design phase of the switching power supply. Usually, many different systems are intimately wired together on the power system and a failure in one system can cascade to other systems.

Undervoltage Conditions

These conditions result from a failure in the power source or distribution system. In a bulk AC-to-DC power supply bus system, undervoltage can be caused by a failure in the supply itself or a load failure causing the main bulk supply to go into foldback. In a generator (alternator)–battery system a generator failure causes the system to only run from the battery. This is a degenerating condition, and the system voltage will only worsen. The power supply designer must decide how much below the specified minimum input operating voltage the power supply and load circuitry must work. In bulk power systems, it may be necessary only to inhibit operation at the minimum specified voltage value. But in generator (alternator)–battery systems, the operator's life may depend on your system's operation for as long as possible, no matter what the conditions. In this situation, the amount of margin designed into the supply as far as low input line operation may have to be significant. This type of situation will affect the design of the supply, such as the type of foldback used, the selection of power switches, and the amount of heatsinking required. At abnormally low input voltages, the amount of average current required by the supply increases significantly. This makes the power switch and transformer heat up beyond normal levels, and overdissipation is the consideration. Typically one can allow operation until just short of the point of destruction, which is dictated by the thermal limits of the power switch and the transformer and drive–IC supply voltages. Operation should be inhibited below this point since the unit will destroy itself anyway.

Overvoltage Conditions

This situation arises when a failure in the power source occurs. In the bulk power source system, the bulk supply has lost its voltage feedback loop. In the alternator–battery system, the alternator has lost its field line or the regulator has failed. This type of situation is indicated by a long-term voltage that exceeds the normal maximum input voltage specification. The designer should become acquainted with the power source failure characteristics and design accordingly. First, overrate the voltage of the semiconductors in the supply. Second, design a scheme of providing voltage to the control IC and power switch that is relatively immune to the input voltage, such as using the secondary voltages, or have a linear regulator hooked to the input line that has a high input voltage rating. Because this can be a long-term adverse condition, components such as transorbs, which are designed for instantaneous surges, cannot be of any help.

Line Dropout

This condition is caused by relays that switch power sources or batteries. Line dropouts are typical in aircraft power buses and UPS backed-up power systems. It is a sudden disappearance of the input voltage for a short period of time. In order to design around such a condition it is necessary to increase the value of the bulk input filter capacitor to the point where the supply can continue to maintain its outputs within the specified range for the worst-case dropout period. Also, a series rectifier must be added between the input line and the bulk filter capacitor to prevent losing its charge back into the line during the dropout.

Surges

Surges and transients occur predominantly in generator (alternator)–battery-type DC systems. This occurs when a heavy load turns off and the alternator regulator cannot respond quickly enough. In the automotive industry this is called "load dump". The surge condition can last for many milliseconds and go as high as 5 to 6 times the bus voltage specification. This condition is just within the reach of what transorbs can handle if a series impedance is added between the line and the transorb. This typically can be an EMI filter choke. A good design practice in this situation is to select parts that are directly connected to the input line and that have maximum rated breakdown voltages greater that the worst anticipated voltage surge.

8.2 Protecting the Load from the Supply and Itself

Protection of the load circuitry should be given a high priority in the design of a switching power supply since it is usually much more expensive than the power supply itself. This means that the supply designer must provide a means to detect and counteract the effects of a failure within the supply or on its outputs. To adequately address this facet of the design, the designer must become acquainted with the true destructive limits of all the loads and the nature of the possible failure modes within the chosen switching power supply architecture. Then the designer must investigate the repair philosophy of the market this product is entering. Only when all the above considerations are addressed can the designer design a switching supply that will meet the needs of the system.

The types of failure that can occur on the output lines of the switching power supply are basically overvoltage, overcurrent, and short circuits. Overvoltage conditions are typically caused when the power supply loses its voltage sense loop or a circuit failure causes a higher voltage to short-circuit to a lower voltage. An overcurrent condition is typically caused by a short circuit in the load that is in series with a low resistance. A short circuit is an overcurrent condition without the current limiting provided by any series resistance.

Hardware implementations of protection circuits to counteract these failure modes fall into three basic categories of operation and maintenance.

1. *Repair and replacement.* This is a protection philosophy of placing sacrificial components within the supply that during an occurrence of failure will fail in such a way as to save the supply or load from destruction. The unit must be repaired after a failure, which means downtime for the unit. These include fuses, zener diodes, and fusistors.

2. *Protection circuit activation on failure and deactivation on removal of the failure.* These are circuits within the supply that sense the presence of a failure and override the normal functions of the supply such that they reduce the destruction within the failure. They are temporary in nature and are active only during the presence of a failure. They cease function when the failure condition is removed. These include overcurrent foldback circuits and overvoltage override circuits.

3. *Power shutdown and recycling methods.* These circuits, upon sensing a failure, shut off the power supply and the operator must turn off and turn back on the main power switch to restore the unit's operation.

In good power supply designs, these protection schemes are cascaded in order to offer redundant protection to the load in the event of a primary protection circuit failure. The effects of cascading these functions are multiplicative; that is, one obtains much more than twice the protective effects for cascading two circuits. It also gives the entire product a much better degree of graceful degradation under failure conditions.

To properly select the primary method of protection and to design the protection circuit(s), the designer should consider the usage and repair philosophy of the market into which the product is being sold. For instance, if the equipment should never operate during a failure where erroneous operation may result, a repair-after-failure method of protection should be used. The unit would be returned to the repair shop, repaired, and calibrated. If the unit has loads that hangup or stall but can clear themselves, the temporary overcurrent foldback method or power recycling method should be used. This allows the failure to clear itself or the operator to clear the failure, respectively. Usually for the latter two types of protection methods, the cascaded protection method is a fuse that acts as the final protection when all else fails.

8.2.1 Hardware Implementations to Address Overvoltage

Zener Diode

The zener diode is one of the most inexpensive forms of overvoltage protection one can use. The typical zener diode can withstand 10 to 20 times its maximum current over a short period of time without failure. So it can act as an effective transient clamp. Also, when the zeners finally do fail, they fail shortcircuited. The final result is the load never sees a lethal level of supply voltage and the replacement cost of a zener is much less than the load circuit itself. The only shortcomings are the tolerances of typical zener diodes. One may have to buy a tighter tolerance than the standard part so that it will not conduct during normal operation or start conducting after a lethal voltage has reached the load. (See aso Fig. 8.1.)

Figure 8.1

The overvoltage zener diode clamp.

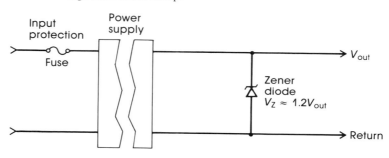

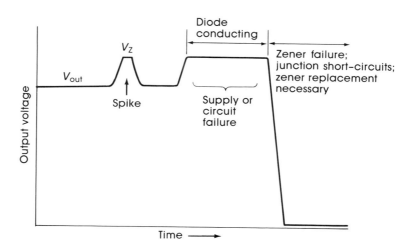

The Overvoltage Crowbar

This is basically a voltage comparator followed by an SCR. The voltage comparator senses the output voltage and activates when a predetermined threshold has been exceeded. The SCR then latches "on," which causes the output to short-circuit to ground, and the supply enters an overcurrent foldback mode of operation. The SCR does not unlatch until all current from the supply disappears. This usually means that either a fuse blows or the supply is turned off. The shortcomings of this method are the threshold accuracy and the gain of the voltage comparator circuit and its sensitivity to system noise, causing false and inaccurate triggering. (See also Fig. 8.2.)

Figure 8.2

The overvoltage crowbar.

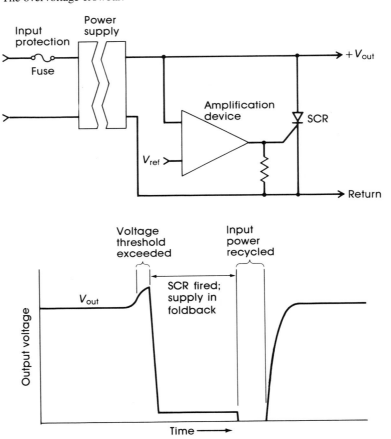

8.2.2 Hardware Implementations to Address Overcurrent

Current Limiting and Foldback

There are several forms of this type of protection circuit, each exhibiting a different characteristic (see Fig. 8.3). It is highly recommended that some form of overcurrent protection be included in every power supply design.

The most rudimentary form of protection is *constant-power limiting*. This is accomplished when the current in the primary of the transformer is sensed and regulated to a constant value when the loads draw too much current. This yields a constant input power, and hence constant output power. The only shortcoming is that it allows increasing current to flow into a failure as the short circuit becomes more severe. This could result in burning of the PCB during failures. (See also Fig. 8.3.)

Constant current limiting is a method for monitoring the load current. When excessive load current is drawn, it decreases the output voltage such that the output current is held constant. (See also Fig. 8.4.)

Overcurrent foldback is a method that provides the best degree of load protection during an overcurrent failure. Both the output current and voltage are reduced so that the power entering a short circuit is greatly reduced. (See also Fig. 8.5.)

The method of overcurrent shutdown turns the unit off when an over-current failure occurs. It also requires that the supply has output current

Figure 8.3

Output curves for different types of overcurrent limiting.

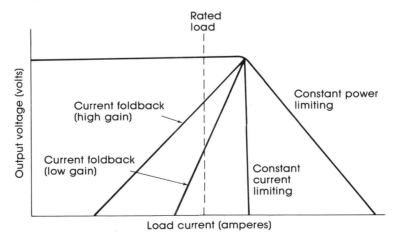

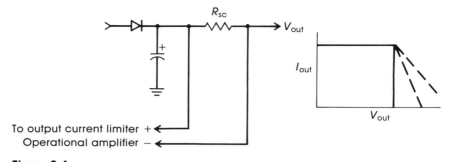

Figure 8.4

Constant current limiting in overcurrent protection sensing.

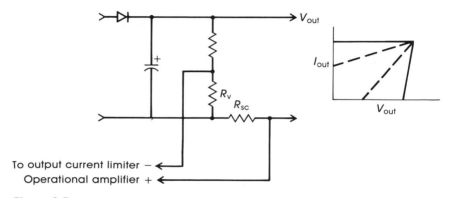

Figure 8.5

Current foldback limiting in overcurrent protection sensing.

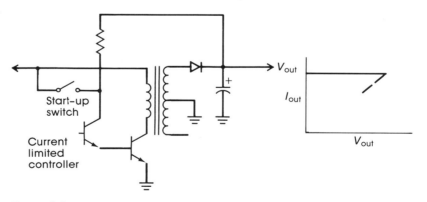

Figure 8.6

Overcurrent shutdown in overcurrent protection sensing.

limiting as part of its operation. At powerup, a startup circuit allows current to be passed from the input line to the control IC and power switch driver sections. After the supply reaches its rated output voltage, the startup circuit is turned off and the supply now derives its control and drive voltages from its own outputs. When an overcurrent failure occurs, the overcurrent protection circuit reduces the outputs. This, in turn, reduces the control and drive voltages, the output voltages, and so on. In short, this quickly produces a degenerative condition and the supply completely turns off within a couple of cycles. To restore the supply to operation, the input power must be turned off and turned back on in order to reactivate the startup (or bootstrap) circuit. (See also Fig. 8.6.)

9

Miscellaneous Topics

9.1 Power Supply and System Grounds

The design of the ground layout within the switching power supply is important to ensure not only its stable operation but also long-term reliability. The power supply engineer, who probably will understand the power supply problems better than any other engineer involved with the product, should also be attentive to the design of the entire power distribution layout within the end product. The laying out of the printed circuit board(s) for the power supply and the product is where a little trust in your fellow man can get you in a lot of trouble. To the majority of draftspersons and all the computer-aided design (CAD) programs, a ground is any line that measures zero "DC" volts and acts as an infinite current sink to any amount of current that is dumped into it without affecting its infinite sinking characteristics. Nothing could be further from the truth.

Grounds fulfill two electrical functions: (1) return the current used by the circuitry back to the power source and (2) couple associated circuits together. In the case of switching power supplies, the current generated by the circuit has a wealth of high-frequency spectral components and at high current levels. These high-frequency current components make any small inductances associated with wires and PCB traces become significant in their ability to turn current into voltage. These equivalent inductances are in series with the current path and become greater in proportion to the distance the current is forced to travel. Ground planes are not immune to the effects of series parasitic inductances. A ground plane resembles a matrix of equal inductances in every direction from any point within the ground plane. These equivalent series inductances are much smaller in value than those associated with a single PCB trace but nonetheless can present a problem when low-level control grounds are mixed. High current with high-frequency components will tend to take the path of least resistance and inductance, which is typically the most direct path within the ground plane.

In PWM switching power supplies, ground-related problems will usually manifest themselves in three design areas: ripple voltage and AC current sharing of paralleled capacitors, closed-loop stability of the supply, and EMI/RFI levels. For capacitors, the PCB traces add to the equivalent series inductance (ESL) and equivalent series resistance (ESR) that is already present in the capacitor structure. This isolates the actual storage capability of the capacitor from the signal to be filtered, hence reducing the effectiveness of the capacitor. When large input or output filter capacitors are electrically paralleled, the capacitor with the shortest pathlength dissipates more power internally than do the other capacitors because it receives the largest value of root-mean-square (RMS) ripple current. Grounds associated with the control section of the switching power supply must be kept as quiet as possible since the amplifiers are sensitive to millivolt changes on their inputs. If the low-current grounds associated with the controller amplifiers are mixed with the high-current grounds in the power sections of the supply, the voltages generated by the parasitic inductances within the high-current areas are arithmetically added to the feedback voltage or current signals. The amplifiers are sensitive to changes in voltage on the order of 10 mV, so it doesn't take much current to affect the operation of the amplifier. When the control and power grounds are mixed, stabilization of the voltage and current loops becomes more difficult, if not impossible. The control ground should be connected directly to the output terminals in nonisolated applications, since the output driver ground section of the controller is much more tolerant of noise. Finally, good ground design goes hand-in-hand with good RFI/EMI design practices. Current causes RFI, not voltage, and the longer that current must travel, the greater the radiated power of the noise. Please refer to Section 9.3 for RFI/EMI considerations.

There are three basic grounds within a minimum switching power supply (see Fig. 9.1). The first is the return path for the primary winding and the power switch. The current waveform that runs through this return ground path is identical to the transformer's primary current waveform, including the high peak current and the rapid rise and fall times. Logically, the ground side of the power switch should be connected directly to the ground side of the bulk input filter capacitor,

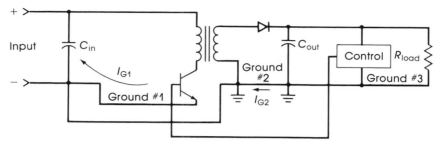

Figure 9.1
Power supply grounds.

which is the instantaneous source of the primary current. The second ground is the transformer's secondary output ground lead, which is included in the loop of the secondary winding, the rectifier, and the output filter capacitor. This ground runs between the center-tap of the secondary winding to the ground side of the output filter capacitor. Its current is identical to the secondary winding's current, once again including the high peak currents and fast rise and fall times. The ground side of the output filter capacitor(s) should be wired directly to the output ground lead of the secondary winding or full-wave bridge. The third ground is the control ground, which should include all the low-level circuitry surrounding the control IC. All ground traces from this section should be at the end of a branch in a "one-point ground" arrangement. That is, dedicated ground traces should be run to the voltage output terminal for the error amplifiers and to the power switch for the output driver. No high currents from the power sections of the supply should be permitted to flow along these traces. One-point grounding is the design philosophy behind the PCB layout within a switching power supply.

Within the entire product (Fig. 9.2), one-point grounding is also important. All filter capacitors should have good high- and low-frequency filtering characteristics. By intelligently placing the main system filtering capacitors within the power distribution system, one can effectively restrict the high-frequency noise to the local areas in which they are generated. This greatly reduces RFI and circuit interaction. The filter capacitors can source the high-frequency current transients such as those generated by logic circuits, and the energy removed by the transients is then replaced by the DC current provided by the power supply. This allows current with only low-frequency components to be carried by the long lengths of system power wiring. It also prevents the high-frequency transients from leaving the "noisy" circuit and entering a "quiet" cir-

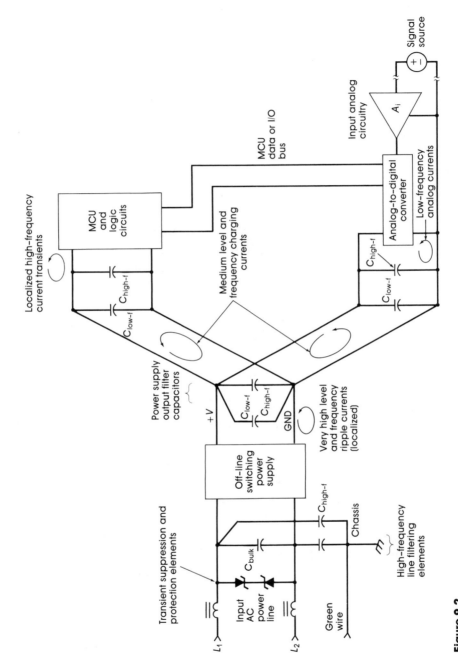

Figure 9.2

Recommended ground layout for a typical system.

cuit such as an analog section. One-point power distribution will not be a popular approach among those who lay out the PCBs, but it does provide the optimal results.

9.2 The Use and Design of Clamps and Snubbers

The use of clamps and snubbers within switching power supplies is intended to reduce the lethal effects of voltage spikes generated by the supply itself and to reduce RFI/EMI emissions. The amplitude and shape of these spikes can violate the limits imposed by the forward-biased safe operating area (FBSOA) and reverse-biased safe operating area (RBSOA) of the power switches used within the switching power supply. The rapid voltage and current transitions associated with a voltage spike generate a wealth of high-frequency RF components that are easily radiated to the environment. It must be stated, though, that the addition of a snubber or clamp to the power supply should be a path of last resort for the designer. Much can be done by the designer during the transformer design stage, the component selection stage, and PCB layout stage to reduce, if not eliminate, the need for a snubber or clamp. Actually, the appropriate time to design the final clamp or snubber is after the first production unit is built. The physical shape and energy within the voltage spikes are completely determined by the parasitic inductances and capacitances contributed by the physical layout and transformer design. These are typically not finalized until just before the production run. The designer should not introduce an additional loss within the supply unless it is absolutely necessary.

Many mistakes are made by the designers of switching power supplies as to when to use a clamp or a snubber and also, if using a snubber, how to determine the optimum values of the resistor and capacitor. Although both the clamp and the snubber reduce the peak voltage of a spike, they are intended to accomplish two different purposes. First, the designer must keep in mind what is to be protected. Typically it is the semiconductor power switch that can be a bipolar power transistor or a power MOSFET. Power transistors can more readily handle avalanche breakdown (overvoltage) than can power MOSFETs, which are more sensitive to avalanche breakdown. Power transistors, though, have a significant second breakdown and current-crowding problem that occurs below their rated avalanche voltage (V_{ceo}). So which does one use for a particular situation? Well, the answer is easy and not so easy. It is all dictated by the SOA curves of the power switch. That's the easy answer. The difficulty is presented when one tries to plot the instantaneous turn-on and turn-off load lines on the SOA curves. The designer must get a stable picture of both the collector (or drain) voltage and current and pick off the values of the instantaneous voltage and current at the same instances in time and then plot these V and I values on the appropriate SOA curve. For transistors the RBSOA curve is used for turn-off and the FBSOA is used for turn-on. For MOSFETs, the SSOA curve is used for turn-off and the FBSOA is used for turn-on. If any point on the plotted curve falls outside the SOA curves, the switch is bound to fail. The failure may not occur immediately or even in the first 100 units built by production, but somewhere, sometime a unit will fail for this reason. So the designer should carefully observe the current and voltage switching waveforms in consideration of the power switch SOA curves and pursue the appropriate actions within the design.

The classic diode and zener clamps (Fig. 9.3) are used exclusively for an avalanche (or overvoltage) breakdown failure condition. This is when the voltage spike exceeds the V_{ceo} (transistor) or V_{dss} (MOSFET)

Figure 9.3

Hard voltage clamps.

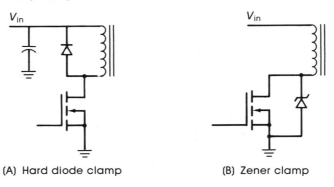

(A) Hard diode clamp   (B) Zener clamp

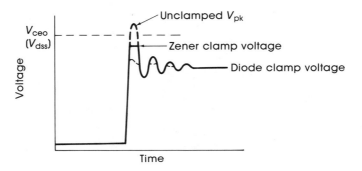

limits of the power switch *without* violating any second breakdown or current-crowding limits of the power transistor, if used. If a rectifier diode clamp is used, the diode should be an ultra-fast-recovery diode. This is because the forward recovery time (turn-on time) of the ultra-fast-revovery diode is the fastest of all the *P–N* junction rectifiers. This turn-on time is important in order to begin clamping the spike as soon as possible. Slower diodes will allow the spike voltage to rise above the designed clamping voltage before the diode begins to conduct the spike's energy to a low-impedance "sink." All zener diodes begin conducting very quickly, but do exhibit a series resistance between the spike and the "zener" portion of its equivalent circuit. This can allow the spike voltage to rise slightly higher than the rated zener voltage. How much higher is dependent on the energy contained within the spike and the dynamic impedance (Z_{zt}) of the zener diode. Rectifier diode clamps are typically returned to the input bulk filter capacitor in order to recover most of the energy within the spike. For zener clamps, though, the spike's energy is lost in the $E \times I$ product experienced across the zener.

One variation on the diode clamp is what could be called "soft clamps" (Fig. 9.4). These clamps return the spike's energy to a "soft"

Figure 9.4

Soft voltage clamps.

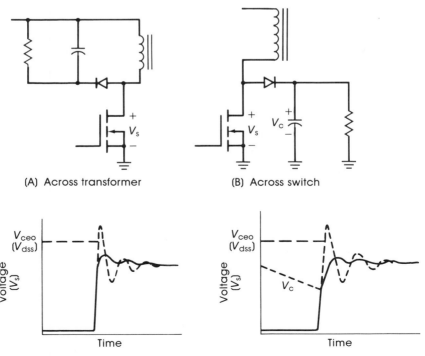

(A) Across transformer   (B) Across switch

current sink such as a medium-sized capacitor. To use them, the designer must know the energy within the spike in order to choose the optimum capacitor size. The capacitor voltage is then bled off between the occurrences of the spikes. This method not only clamps the spike's voltage but can also do some shaping of the spike's leading edge, which can aid in avoiding the second breakdown and current-crowding conditions of the power transistors. These are tricky to design since the "cut-in" voltage and the peak clamping voltage can be variable with the duty cycle and input voltage. The time constants of both the charging and discharging of the capacitor must be carefully evaluated for the entire dynamic range of the power supply. Also, since these employ a diode, its function is intended for a rising or falling edge transition only. Needless to say, the diode must once again be an ultra-fast-recovery diode.

One technique that can be used in flyback supplies is called a *clamp winding* (Fig. 9.5). Since the flyback supply uses its transformer for energy storage, and the time between the power switch turning off and the output rectifier turning on leaves the core completely unloaded, the self-generated voltage spikes can be quite severe. The clamp winding is a winding whose turns ratio is 1:1 with the primary winding. Its purpose is to conduct only during unloaded core periods and return its energy to the input filter capacitor. To be effective, the clamp winding must be tightly bifilar-wound with the primary winding in order to not only clamp the unloaded core energy but also capacitively couple to the primary winding's leakage inductance. If it is not tightly coupled, the clamp winding will act only on the core's energy and not the primary leakage inductance, resulting in a spike on the primary caused by the leakage inductance. In that case an external clamp or snubber would still have to be added. The clamp winding is particularly effective in transformers requiring a high dielectric strength between the primary and the secondary. In this case, the coupling between the primary and

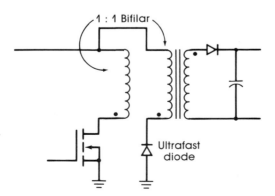

Figure 9.5
Clamp winding for flyback regulators.

the secondary is poor and the tightly coupled clamp winding can act quickly to get rid of the energy associated with the primary leakage inductance and the instantaneously unloaded core. Since the clamp winding returns the energy to the input, it does not add a significant loss to the supply. So it should be investigated before the use of a snubber is considered within a flyback power supply.

Lastly, we consider the snubber (Fig. 9.6). Its purpose is to control both the spike's transition rate and shape and the peak voltage. It is also the most effective means of avoiding the secondary breakdown and current-crowding failure modes of the bipolar power transistors. Its theory of operation is to make a tuned circuit in conjunction with the

Figure 9.6

Snubbers.

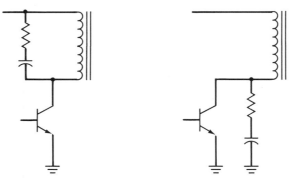

(A) Snubber across transformer (parallel snubbing)

(B) Snubber across switch (series snubbing)

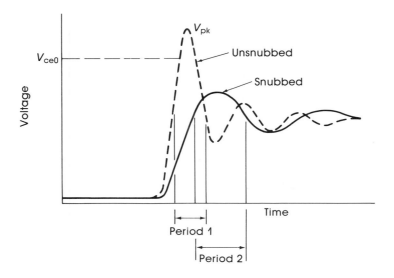

parasitic inductances and capacitances of the transformer and the physical PCB layout and to dissipate that tuned energy within the resistor. In short, the snubber and parasitic reactances make a "lossy" tank circuit. Since the parasitic capacitances and inductances are distributed throughout the surrounding components and layout, it is difficult to locate their sources and measure their values, so modeling is impractical. The designer instead works with the final "production" layout and treats parasitic reactances as a "blackbox" circuit where the reactances are lumped into one or two reactive elements. There are no easy equations to define the values of the resistor and capacitor within the snubber. Instead, a procedure (Carston, 1986b) that seems to work quite well is the following.

1. Place an oscilloscope voltage probe across the element that is to be snubbed. Take care to minimize any stray pickup that might affect the waveshape by placing the ground lead very close to the element. Then record the peak voltage and the ringing frequency (F_o) of the spike waveform.

2. Place a very small capacitor in parallel with the element terminals to be snubbed. Increase the value of the capacitor until the ringing frequency is cut to half of the original unsnubbed ringing frequency. At this point, the value of the sum of the parasitic capacitances (C_o) will be one-third of the paralleled capacitor value.

3. Calculate the estimated optimum value of the snubber's damping resistor by

$$R = \frac{F_o \times C_o}{6.28}$$

4. Place this value resistor in series with the added capacitor. This resistor value may have to be varied one way or the other to yield the desired peak voltage and damping.

Note: The power dissipated within the snubber is given by $P_s = CF \cdot (V_{p-p})^2$, where C is the snubber capacitance, F the power supply's operating frequency, and V_{p-p} the peak-to-peak voltage across the snubber capacitor. Although this approach yields very good results for the snubber values, it may present a problem in power dissipation. A higher value of resistor and a smaller value of capacitor will decrease the power within the snubber, but the damping factor will worsen and the peak voltage of the spike will grow. The designer must trade off the dissipated power and the peak spike voltage with the purpose of the addition of the snubber.

The use of clamps or snubbers should be a last resort. It is important that the designer be aware of what other areas within the design would be more effective in combatting voltage spikes and RFI generation than adding a clamp or a snubber. They can be viewed as "design patches" for a design weakness or a physical shortcoming of a component. If one is needed, though, the designer should be aware of the factors that indicate the optimum choice and design.

9.3 RFI and EMI Design Considerations

Radiofrequency interference and electromagnetic interference present a significant problem within PWM switching power supplies. The very rapid transitions present on the high current waveforms within the PWM switching power supply are the primary sources of interference. Modern designs of PWM switching power supplies are reaching operating frequencies of 500 kHz to 1 MHz, which makes it easier to radiate interference to the surrounding environment. Significant spectral components of tens to hundreds of megahertz are present and are enhanced the faster the power switches are forced to switch.

Controlling the emission of RFI and EMI is probably the most "blackbox" art within the realm of designing switching power supplies. The designer is not certain about possessing a mechanical and electrical design that will pass the specified requirements until the product is actually tested. This no doubt worries the designer during the entire design process. Each electrical and physical design of a switching power supply has its unique filtering and shielding solution. Frequently the shielding of the RFI generated by the supply itself can be accomplished by product enclosure through the use of a conductive paint or metallic enclosure. For the most part, the control of the RFI and EMI of the product is left in the hands of the switching power supply designer. Controlling RFI is done through the use of ground planes, shields, filters, snubbers, and intelligent PCB layout.

The first place to start for controlling RFI and EMI is the design of the switching regulator itself. First, all high current traces should be as short as possible, with power switches, diodes, transformers, and capacitors being as physically close to one another as possible. Second, the designer should observe all switching transitions within the supply and check whether there are spikes on the waveforms or whether the waveform has a very high rate of change (dv/dt or di/dt). If so, perhaps a snubber is called for. There should be ample ground planes for conducting the high currents. Remember, "an ounce of prevention is worth a pound of cure."

Shields are designed for trapping any RF energy within an enclosure. This feature is based on the Gaussian sphere theory; specifically the shield must enclose the entire RF-generating circuit and form a "shorted (short-circuited) turn" around it. Any joints between sections responsible for shielding should have a very low contact resistance, or else the effectiveness of the shield will be greatly impaired. Single plates of shield provide little or absolutely no shielding for RF; it must completely surround the supply or product.

Filtering takes the form of RFI filter chokes in series with all wires that enter or exit the power supply. Shunt high-frequency capacitors are also used to short-circuit the conducted RF energy to a large ground plane and to the green wire ground lead back to "Earth." RFI filtering should take place where the wires physically enter or exit the enclosure. This prevents the wire from radiating outside noise into the enclosure or picking up noise from inside and radiating it outside.

The limits of RF radiation are different for each region of the world; this is because of the allocation of the radio spectrum for each section. In areas of the spectrum where broadcast or radio communications are allocated, the limit for the radiated RF power is lower. The regulations for each region of the world are generally governed by the same approval bodies that govern safety, except within the United States where they are governed by the Federal Communications Commission (FCC). Testing of a product not only takes the form of measuring the radiated

Figure 9.7
Conducted EMI limits (VDE).

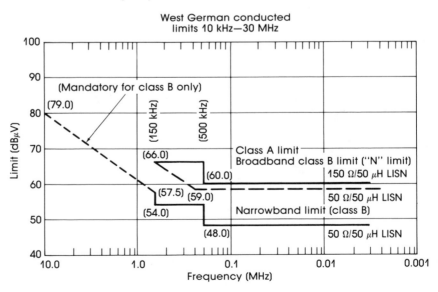

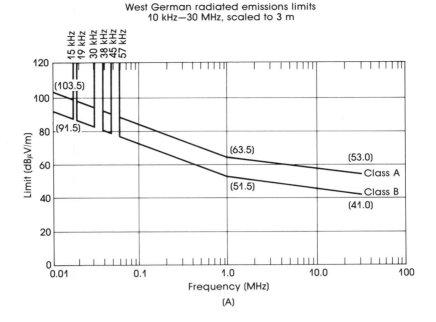

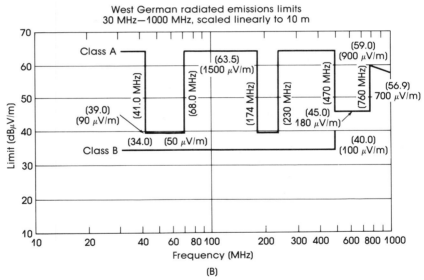

Figure 9.8

Radiated RFI limits (VDE).

RF field strength around the unit but also the RF conducted through each wire connected to the unit. All these tests view the RF spectrum from 10 kHz through hundreds of megahertz. So the testing is quite thorough. The spectral limits for Europe can be seen in Figures 9.7 and 9.8.

The governing bodies have the authority to seize and impound equipment and property as well as impose fines if any product is found to not conform to their specifications. So don't take this testing too lightly.

9.4 Power Supply and Product Safety Considerations

Each industrialized country has adopted safety standards for assurance that the products marketed within their borders are safe for the end user and repair personnel. For electronic equipment, the term *safe* means that the product must not present a shock, fire, energy, or mechanical hazard to the people using it and provide a reasonable degree of safety to those persons repairing it. Therefore, if it is your company's intent to market the product anywhere in the world, safety must be a major concern to the designer of the power supply system. For the power supply designer, there is a double safety requirement. The power supply must meet the safety requirements not only for itself but also for the entire product when working within the product. This means that the power supply designer must become involved in the product design as well. This facet of the design can be viewed as a headache to the designer because it restricts the choice of components and influences the physical design of the transformers, PCB, and chassis, which can hinder any effort to minimize the size and cost of the power supply.

Regulatory agencies have been established in the major industrialized countries to create and maintain the safety standards and to test new products entering their market for compliance with their standards. Until recently, the differences between each regulatory agency's standards were sufficient as to require reapproval within each agency's jurisdiction. This was a major headache to manufacturers who desired to establish an international market for their products. Fortunately, especially within the European Economic Community (EEC), efforts to adopt overall standards have succeeded. Although there are still separate regulatory agencies within each country in Europe, they now recognize the approval of another agency within their respective country. Care must still be taken in Europe to assure oneself that there is no requirement that goes beyond the requirements of the joint agreements. Within North America, similar cooperation exists between the United States and Canada, and the regulatory agencies tend to recognize the standard that has a more stringent set of tests. It is highly advisable to obtain copies of the applicable standards early in the design process and talk to either the regulatory agency(s) or consulting firm whose job is to help companies through the approval process.

The safety regulatory agencies within North America are as follows: in the United States, the Underwriters Laboratory (UL); and in Canada, the Canadian Standards Asociation (CSA). In the Common Market countries, the joint standards are maintained by the International Electrotechnical Commission, but they do not physically perform the testing of the products. There are about a dozen regulatory agencies throughout Europe and the Common Market countries, but the major agencies are VDE (West Germany) and British Standards Institution (BSI). Once again, there are companies that are authorized to perform the compliance testing for the overseas regulatory agencies.

The safety regulatory agencies have categories of products that each have their own standards. The title of the standard under which your company's product will be tested will probably not match the end purpose of the product but historically has been grouped under that category. For power supplies and end products the applicable standards for consumer product categories and some industrial product categories are as listed in Table 9.1.

Table 9.1

Applicable Safety Standards for Power Supplies and Transformers

UL (United States)	
UL 506	Specialty transformers
UL 1012	Power supplies
UL 1411	Audio, radio, and TV transformers
UL 1585	Class 2 and class 3 transformers
UL 1310	Direct plug-in units
UL 478	Electronic data processing equipment
UL 1459	Telephone equipment
VDE (West Germany)	
VDE 0804/1.83	Telecom equipment
VDE 0805/11.83	Electronic data processing equipment
VDE 0806/8.81	Office machines
CSA (Canada)	
C22.2 0.7-M1985	Equipment connected to telecom networks
C22.2 No. 143-1975	Office machines
C22.2 No. 154-M1983	Electronic data processing equipment
IEC (International)	
IEC 380 (1985)	Office business machines
IEC 435 (1983)	EDP
IEC Guide 105	Telecom equipment

In an effort to give the designer a feeling of what to expect, a brief summary of some of the major design-related requirements that directly affect the design of a switching power supply follows.

1. *Marking and labeling.* The unit must be marked with the input ratings (voltage and current), the type of power (AC or DC), the line frequency, the insulation class, and the rated load (if it is only a power supply being marketed). The labels must strictly adhere to the requirements of the standard, including any caution statements that may be applicable.

2. *Clearance and creepage.* This is the minimum distance between conductors that are either the opposite phase of the input line(s) and an isolated output line. Clearance is the distance with only air acting as an insulator between the conductors. Creepage is the distance along the surface of an insulator. For a 250-V_{RMS} input, the clearance and creepage requirement is for a minimum of 3 mm between opposite phases of the input line. The clearance and creepage requirement between the input line circuits and the case or output(s) should be 8 mm. This requirement can be lessened if an approved insulator is placed between the two conductors.

3. *Nonhazardous voltages.* These are voltages that are not considered dangerous if a grounded person were to come in contact with them. These are any conductors that do not exceed a peak voltage of 42.4 V and have a current-carrying capacity of less than 8 A peak. These safety extra-low voltages (SELVs) must be physically separated and insulated from voltages that are considered hazardous.

4. *Operator Access.* The operator should not inadvertently come in contact with hazardous voltages from outside the product. Access to such voltages must be allowed only by the use of a "tool." The operator must consciously accept the risk of injury by deciding to use a tool to open the product. Some circuits may even be required to have an interlock to cut off the hazardous voltages even after "tool" access has been made.

5. *Wiring harnesses.* All wires within a harness must have an insulation rating that meets or exceeds that rating required for the highest voltage within the harness.

6. *Enclosures.* The enclosure provides the first level of protection. It must either be constructed of an insulator, or if it is metal, grounded to earth ground (green or green–yellow wire). All metallic surfaces

or controls must also be similarly grounded. The maximum series resistance from any exposed metal surface and the earth ground terminal will not exceed 0.1 (one-tenth) Ω when measured by passing 25 A from the surface to the earth ground. The earth ground wire within the AC cord must be larger than #18 AWG (American wire gauge).

7. *Resistance to fire*. Materials used within the supply and product must not act as fuel for sustaining a fire. Underwriters Laboratory UL-478 flame ratings are stricter that those of the European community, so the designer is safe in selecting 94-V2- or 94-HF2-rated component materials or higher. The PCBs should have similar rating and be marked as such.

8. *Insulation resistance and dielectric strength*. This is a series of tests designed to measure the ability of the product and insulators to withstand abnormal operating conditions such as lightning strikes and power surges. These requirements are written in terms of AC voltages. These can be lethal to any linear or switching regulator for reasons not intended by the test. For these tests, the input lines are short-circuited together, and the outputs are separately short-circuited (i.e., the output line with its own respective return line). A high-voltage AC voltage is placed between input, output, and chassis separately, and the leakage current is monitored. Unfortunately, the transformer windings act like a small AC capacitor that allows a small AC current to flow through the electronic components. This will not only build up lethal voltages across the semiconductors, but their AC voltages will swing in both polarities. Regulatory agencies do appreciate this fact and usually allow the test to be conducted using the peak DC equivalent to the specified AC values. Also, any RFI/EMI components connected between the input line and ground or outputs must be able to withstand the test voltages. The leakage current measured during the test is then used to calculate the dielectric resistance of the product. In general, a dielectric resistance of 7 $M\Omega$ is required. If the current exceeds a specified maximum value or if an insulator arcs over, the test is failed. The test voltage is applied for one minute. For DC test voltages, it is generally permitted to ramp the voltage up and down in a 2-sec period. (*Note:* Don't ramp the DC voltage too quickly since a changing DC voltage acts like AC through the transformer.)

See Table 9.2 for a listing of IEC dielectric strength tests.

Table 9.2

Dielectric Strength Tests of IEC 380

Application of Voltage between	Class I	Class II
Primary and body	1250	3750
Primary and SELV secondaries	3750	3750
Primary and non-SELV secondaries	1250	1250
SELV secondaries and body	Not tested	Not tested
Non-SELV secondaries and body	1250[a]	2500
SELV and non-SELV secondaries	2500	2500

[a]*Note:* Apply 10 times the working voltage, up to 1250. Tests not needed if secondary operates at less than 30 V_{RMS}.

9.5 Testing Power Supply Units

Does your switching power supply meet or exceed the design specifications originally laid down in the initial phase of the program, or does the off-the-shelf power supply meet your product's requirements? This is where a good knowledge of what the power supply parameters mean and how to test for them is useful. The power supply industry has evolved a set of tests and parameters that form the basis of power supply specifications.

9.5.1 Line Regulation

The test for line regulation measures the amount of change in the output voltage in response to a change in the input voltage. This test is conducted with the power supply delivering its rated power (i.e., all outputs delivering rated current). The output voltage is measured (to 0.1 percent minimum accuracy) at three input voltage levels: minimum, nominal, and maximum specified input voltages. The line regulation, given in percent, is determined by

$$\text{Line regulation} = \frac{V_{out}(\text{high}) - V_{out}(\text{low})}{V_{out}(\text{ideal})} (100)$$

where $V_{out}(\text{high})$ and $V_{out}(\text{low})$ indicate the measured output voltages at the maximum and minimum specified input voltages, respectively. (See also Fig. 9.9.)

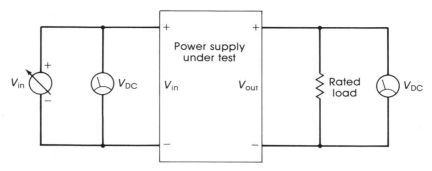

Figure 9.9
Set up for testing line regulation.

9.5.2 Load Regulation

The test for load regulation measures the change in the output voltage in response to a change in the average output load current for each output. This test is conducted with the input voltage set at the nominal voltage (specified operational input voltage). Each output voltage is then measured at 50 percent of rated load current and at 100 percent of rated load current. Load regulation is defined as a percentage and is determined by

$$\text{Load regulation} = \frac{V_{out}(\text{high}) - V_{out}(\text{low})}{V_{out}(\text{ideal})} (100)$$

where $V_{out}(\text{high})$ and $V_{out}(\text{low})$ are the values of the output voltage at 50 percent (low) and 100 percent (high) of the rated output load current, respectively. (See also Fig. 9.10.)

9.5.3 Dynamic Load Response Time

Although this parameter is not usually published for off-the-shelf switching power supplies, it is useful for designers to determine this parameter for their own design. It basically tests for the time it takes for the regulation feedback loop to react to a step change in the output load current and return the output to the specified steady-state voltage. This is one parameter that is worse than that of a linear regulator. Because of the periodic switching nature of the switch-mode power supply, the power supply cannot respond faster than the reciprocal of the frequency of operation of the switching supply. Usually it takes four or more cycles

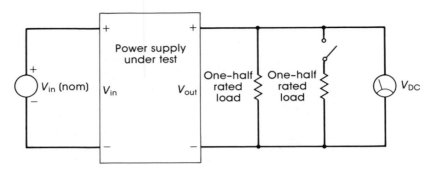

Figure 9.10
Set up for testing load regulation.

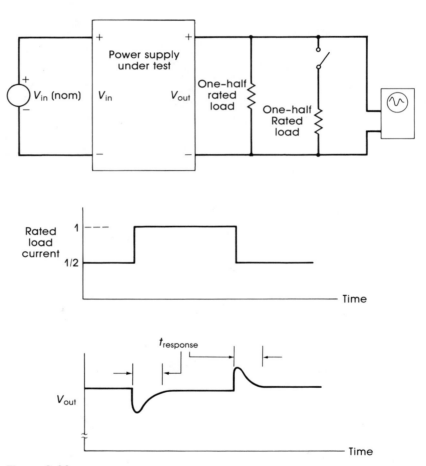

Figure 9.11
Set up for testing dynamic response time.

to replace the energy taken from the output storage elements during the response time and to provide the increased or decreased amount of energy needed by the load at its new load current. The response time and the shape of the response gives the designer an indication of the error amplifier's DC gain and frequency compensation. The more heavily compensated the feedback loop is, the longer the supply's response time and possibly the greater the excursion from the specified output voltage value. (See also Fig. 9.11.)

9.5.4 Dielectric Withstanding Voltage

This is a go/no-go test that checks whether the dielectric isolation between the input, chassis, and output(s) exceeds a specified minimum voltage. The test voltages are typically specified as 50/60-Hz AC voltages, but the peak DC equivalents may by substituted. The overall purpose of this test is to ensure that there is no possibility that potentially lethal voltages from the input line or within the product can reach the end user of the equipment. The particular areas within the product that this test addresses are insulation layers within the power transformer, spacing between isolated traces on the PCB, and input and chassis wiring dielectric strength. A test failure is determined by exceeding a specified current limit during the period when the test voltage is applied to the unit.

Two areas of caution should be addressed in both the application of the dielectric withstanding test and the design of the power supply in order to meet the requirements of the test. First, because the voltages used in this test are many times higher than the maximum voltage ratings of the components, extra care must be taken that lethal voltage gradients not appear across the power supply components. This can be easily overlooked during the test. Foremost, the input and output lines must be short-circuited to their respective return lines. Next, use the peak DC equivalent test voltage when conducting these tests. This is because in the AC test the transformer acts like a small series capacitor between the input and the output. This will cause a small AC current to flow, which is *not* indicative of the dielectric integrity but will develop AC voltages across the power supply components. This causes either outright and/or latent failure of the components as a result of voltage overstress. The second area of caution is that all components that are placed between the input and/or output(s) and the chassis must have voltage ratings greater than the peak test voltage being applied to the

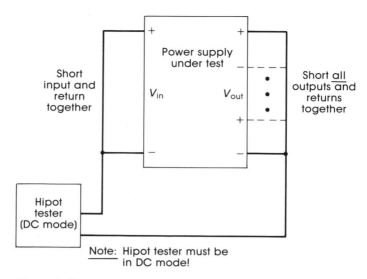

Figure 9.12

Set up for testing dielectric withstanding voltage input to output (Hipot = high-potential test).

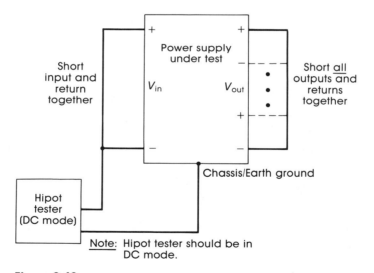

Figure 9.13

Set up for testing dielectric withstanding voltage input to chassis.

power supply. These are typically capacitors included in the RFI/EMI filtering circuits. Any dielectric test that includes the chassis will place the full test voltage across the terminals of these components. A DC dielectric test should also be used in this case since the capacitors will cause a small AC current to flow during an AC test and may inadver-

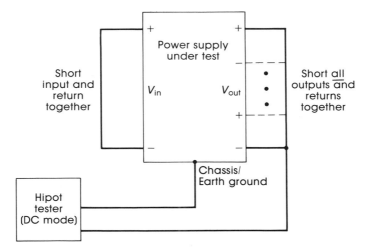

Figure 9.14
Set up for testing dielectric withstanding voltage output to chassis.

tently cause the supply to fail the test. In all cases, the ramping up of the DC test voltage should never be less than 2 sec since a rapid change in voltage would be like applying high-voltage AC to the supply.

Please refer to the appropriate safety specification for your product's market for test requirements and test limits. Remember that all the safety regulatory agencies do accept the use of the peak DC equivalent test voltages in place of any specified AC test voltages. Setups for testing dielectric withstanding voltage are shown in Figures 9.12 to 9.14.

9.5.5 Holdup Time

The test for holdup time (Fig. 9.15) determines the length of time that the output can provide rated power after the removal of the input power. This test gives an indication as to how the entire product will behave during a power interruption. In DC power bus systems, power interruptions are usually caused by the power source being switched by relays, and in AC utility power it may be caused by a heavy surge current being drawn by another load within the local line. In AC systems, a dropout is usually defined as the removal of one-half of a cycle. Within AC and DC power systems, the dropout is typically defined as 8 msec, but this may vary for different applications.

The requirements of this test dictate the amount of energy that has to be stored in the input filter–storage capacitor within the power supply. This test makes the value of the input capacitor many times larger than what is normally required for stable steady-state operation and may pose

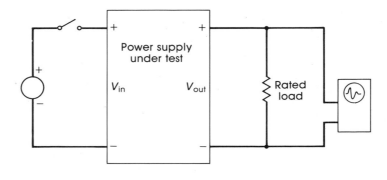

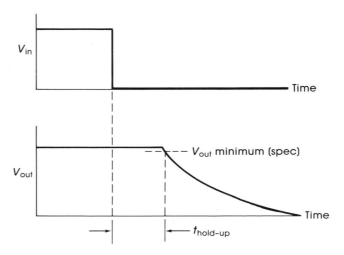

Figure 9.15
Testing for hold up time.

a space problem for space-limited applications. Typical capacitor values to meet this test can be in the range up to thousands of microfarads. Also associated with this requirement is the need for a power-down sensing circuit to alert the load of an imminent loss of power for a "graceful" recovery.

9.5.6 Overcurrent Limit Test

Although the overcurrent limit test is not typically specified in off-the-shelf power supplies, it does offer some valuable information for the power supply designer. The designer can check the overcurrent trip point and the degree of voltage foldback that the supply provides during

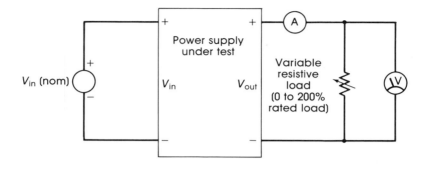

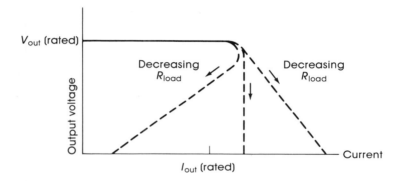

Figure 9.16

Determining the overcurrent foldback characteristics.

an overcurrent condition (see Fig. 9.16). In a multiple-output switching power supply, the response of each output to an overcurrent condition is totally dependent on the method of overcurrent sensing. It is guaranteed that all the outputs will not behave the same.

The test must be conducted with a resistive load and not an active load. An active load simulates a resistive load by regulating the current through a pass transistor so that the $E-I$ relationship is maintained. An active load will "latch up" after the overcurrent trip point is reached in an effort to draw more current than the supply is able to provide. The active load also will tend to interact with the overcurrent control circuit and result in an oscillation. Hence only a variable resistance can be used in plotting the overcurrent characteristic curve.

10

Closing the Loop—Feedback and Stability

It is the desire of all designers of power supplies, whether they are switching or not, for accurate and tight regulation of the output voltage(s). To accomplish this requires a high DC gain. But with high gain comes the possibility of instability. As in all "real-world" control applications with electronic negative feedback loops, the process to be controlled is much slower than the abilities of the electronic feedback path. So the gain and the "responsiveness" of the feedback path must be tailored to the process it is controlling.

Feedback theory can, unfortunately, be difficult to learn and is even easier to forget. Hence the typical approach to compensating the feedback path within switching power supplies is one of trial and error. Sometimes this works with satisfactory results, but it becomes a worry for the designer that sometime during the power supply's operating life a certain combination of input line and load may be encountered in which the supply may become unstable. Fortunately, recent research has resulted in better definitions and models of power supplies. As a result, through the use of some simple tools, the designer can adequately compensate the error amplifier and ensure stability.

10.1 The Bode Plot as a Basic Tool

The Bode plot is a simple and convenient method to represent the response of a circuit over a range of frequencies. It consists of two parameters, gain and phase, which are plotted against frequency. Since the range of the values for the gain can be very large and since the gain function is nonlinear, it is convenient to represent its function as a logarithm. The amplitude level of a signal is represented as decibels above one volt (dBV) or dB with an implied "V" and is calculated by

$$dBV = 20 \log_{10}\left(\frac{V}{1\ V}\right) \qquad (10.1)$$

As one may notice, dB is the log of the ratio of two values. In the case of voltage, the absolute value of gain is referenced to 1.0 V. So 1.0 V is 0 dBV, 2.0 V is +6.02 dBV (+6 dB), and so on. Gain is the relative increase or attenuation of amplitude and is calculated as in Equation (10.1) except the 1.0 V in the denominator is replaced by the initial reference signal voltage level. If the value for voltage doubles, it has increased by +6 dB, or if it halves, it has decreased by −6 dB. Its units are simply in dB (no "V"). Frequency is only plotted on a logarithmic scale to help condense the plot to a more reasonable size.

Before the Bode plot can be used, an input and an output port of the circuit must be defined. This seldom is difficult. The transfer function can then be determined using Kirchhoff's law and defining the components in terms of their AC impedances. Resistors are simply defined as their resistive values, capacitors are given as $1/j\omega C$, and inductors are represented by $j\omega L$. The j in this case is the complex representation of a 90° phase shift of the terminal voltage to the component current. If the j appears in the numerator, it represents a +90° phase shift or a phase lead, or the voltage leads the current such as in an inductor. When the j appears in the denominator, it represents a phase lag or −90° phase shift. In this case the voltage lags the current such as in a capacitor. The omega (ω) represents the frequency dependence of the component value.

Now let's examine some elementary circuits that are particularly approriate in power supplies. First is the simple R–C filter, which can be regarded as a frequency-dependent voltage divider. Its transfer function can be determined by Kirchhoff's law as

$$H = \frac{1/j\omega C}{R + (1/j\omega C)} \qquad (10.2)$$

Reducing this to a more convenient form by multiplying both the numerator and the denominator by $1/j\omega C$, one obtains

$$H = \frac{1}{1 + (R/j\omega C)} \qquad (10.3)$$

To determine the magnitude of the gain over the range of frequencies, one simply omits the j and replaces "ω" with $6.28f$ and multiplies the resulting value by the value of the input voltage function, which is a 1.0

V sine wave at frequency f. To plot the magnitude curve over frequency one simply varies frequency from zero (or very close to zero) to the maximum frequency of interest and calculates the resulting output voltage along the way or

$$V_{out} = H_{(\omega)}V_{in} \qquad (10.4)$$

The plot of the phase is derived by

$$\text{Phase } (f) = \tan^{-1}\left(\frac{\omega C}{R}\right) \qquad (10.5)$$

As one can see when the frequency is zero, the resultant phase shift is also zero. When the frequency is very high (approaching infinity) the phase shift is $-90°$ and is lagging. By once again varying frequency over the same range as in the magnitude plot, one can plot the phase response of the circuit.

To make the plotting of the Bode plot more convenient, several shortcuts are used. First, the frequency at which the magnitude of the two impedances equal one another is called the *corner frequency*. At this point also, half of the circuit's maximum phase shift will be exhibited. The gain and phase curves can be reasonably approximated by replacing them with straight lines called *asymptotes*. This has been done in Figure 10.1. In this case, the first leg, which is below the corner frequency,

Figure 10.1
Frequency characteristics of an R–C filter (integrator).

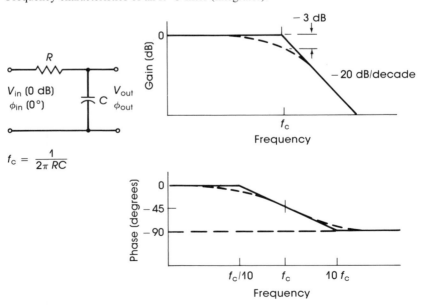

runs from the value of the transfer function at 0 Hz, horizontally to the corner frequency. The second leg has a downward slope from the corner frequency point with a negative slope of -20 dB per decade. The approximation yields an error of only $+3.02$ dB in magnitude. The phase plot begins its negative slope one decade below the corner frequency and ends at its final "high-frequency" phase shift value at one decade above the corner frequency. The point of greatest error between the actual value and the approximated value occurs at the corner frequency. In this case $+/-5.07$ degrees.

The circuit in Figure 10.1 is called a *single-pole* filter. The term "pole" refers to any circuit that causes a decrease in the output voltage (-20 dB/decade) with increasing frequency. These are frequency-dependent terms in the denominator of the transfer function. Zeros are terms that are found in the numerator of the transfer function and cause an increase in the output voltage ($+20$ dB/decade) with increasing frequency such as shown in Figure 10.2. These are referred to as *left-half-plane* poles and zeros, referring to the Nyquist representation of them. Each of these poles contribute a $-90°$ shift (or phase lag), and each zero contributes a $+90°$ phase shift (or phase lead).

When more than one pole or zero is included in the circuit, or when circuits are cascaded, which normally occurs in closed-loop feedback

Figure 10.2

Frequency characteristics of a differentiator.

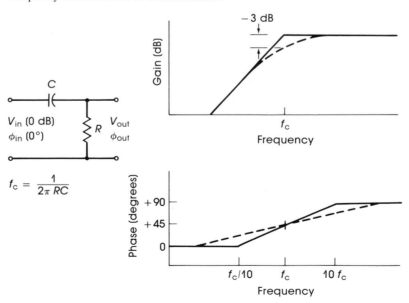

loops, the magnitude slope and phase of each is summed at each frequency. Always start at 0 Hz, since the magnitude response at this point provides the DC amplitude offset for the rest of the amplitude plot. Negative gain slopes will counteract those slopes that are positive, and vice versa. For example, a slope of $+20$ dB/decade when added to a slope of -20 dB/decade will result in a slope of zero dB/decade. Phases simply add at each frequency. In this way the combined overall response can be determined.

10.2 Closing the Loop

Switching power supplies rely on negative feedback to maintain the output voltages at their specified values. To accomplish this, an inverting differential amplifier is used to sense the difference between and ideal voltage (the reference voltage) and the actual output voltage. The inverse of this difference multiplied by the gain of the amplifier results in what is called the *error voltage*. Its role in the power supply is to minimize the error between the reference and the actual output. So as the demands of the load cause the output voltages to rise and fall, the error amplifier changes the energy through the supply to maintain that specified output. If the loads and the input voltage never changed, the gain of the error amplifier would have to be considered only at DC (0 Hz). But that condition never exists. Loads increase and decrease and the input voltage rises and falls. So the error amplifier must respond to these non-DC effects by having gain at higher frequencies.

Any power supply system can be represented by the model in Figure 10.3, where $H_1(s)$ is the AC transfer function of the modulator itself,

Figure 10.3
Closed-loop block diagram.

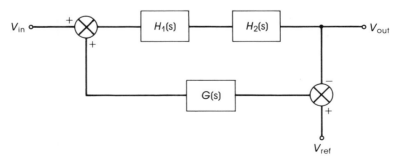

including any transformer; $H_2(s)$ is the transfer function of the output filter; and $G(s)$ is the transfer function of the error amplifier and compensation network.

Each transfer function has its associated gains or attenuations when viewed over a range of frequencies. The gains or attenuations are caused by the variable values of their constituent impedances at different frequencies. The phase behavior is caused by time delays within the circuits.

The closed-loop transfer function can be viewed as

$$\frac{Y(s)}{I_s} = \frac{H_1(s)H_2(s)}{1 + G(s)H_1(s)H_2(s)} \tag{10.6}$$

The poles and zeros of the system can be determined when the actual constituents of the individual blocks within the loop are filled in. Fortunately, the designer does not have to work at this level of mathematics, since the blocks and their behavior have already been determined by our predecessors.

As one proceeds around the loop, the gain (dB) and phase of each successive block are added to the sum prior to inclusion of that block. In the case of switching power supplies, it is convenient to cut the error amplifier out of the closed loop and see what the remaining elements in the loop contribute to the loop's gain and phase characteristics. In this way, the error amplifier can be used to counteract or "compensate" for some of the detrimental effects of the rest of the loop. This partial loop, called the *control-to-output* characteristic, is valuable in designing the error amplifier compensation.

10.3 The Stability Criteria Applied to Power Supplies

The rules for designing a stable power supply are quire simple. The definition of the closed-loop Bode response and the design of the compensation network to meet the stability criteria are the main difficulties.

The rule of stability when applied to power supplies is simply *"Whenever the closed-loop gain is greater than or equal to 1, the closed-loop phase shall never come to within 30° of 360°!"* These should be considered the maximum "boundaries" of the gain and phase Bode plots of the overall closed loop function. Systems that exhibit phase shifts of greater than 330° (360° − 30°) are considered "metastable." These systems will break out into oscillation if they are impinged with a small transient, or will at least ring in an underdamped fashion. For real-world

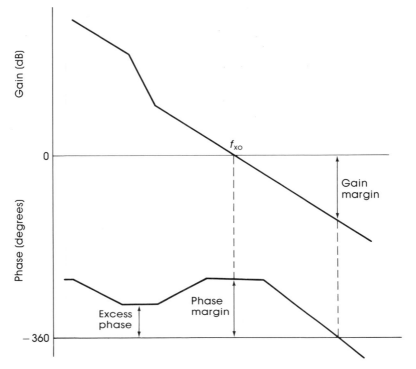

Figure 10.4
Definition of stability terms.

systems, the maximum phase shift should be no greater than 300° to 315° or 45° to 60° away from 360°. Within this range, the supply will respond to transient changes in the operating point in a slightly underdamped to slightly overdamped fashion.

Figure 10.4 shows the meanings of some of the terms used in closed-loop systems:

1. *Phase margin*. This is the value of the phase of the closed-loop transfer function at the gain crossover frequency $[G(s) = 0 \text{ dB}]$.
2. *Gain margin*. This is the value of the gain when the phase of the closed-loop system crosses 360°.
3. *Excess phase*. This is the value of phase for the closed-loop system at its closest point to 360° whenever the closed-loop gain is greater than one. Think of this as the phase margin anywhere within the system bandwidth.

For system stability it is important to *always* keep the phase of the closed-loop system more than the desired excess phase away from 360°!

10.4 The Control-to-Output Transfer Functions of Common Switching Power Supply Topologies

The control-to-output transfer function is a partial model of the final closed-loop, power supply system. It removes the error amplifier from the loop and determines the remaining circuit's inherent gain, poles, and zeros. This allows the designer to concentrate on the method of compensation of the error amplifier in order to make the supply both stable and to meet the load regulation and transient response specifications.

All the common switching power supplies, regardless of their exotic appearance on paper, fall into three basic control-to-output categories. Figure 10.5 shows some of the physical circuit elements within a forward-mode converter that contribute to the low-frequency control-to-output characteristics. There are more items not shown that do contribute poles and zeros to the loop, but their corner frequencies are so very far above the gain crossover frequency that they cannot possibly present a problem to the stability of the system. For forward-mode converters, the process needed to develop the control-to-output Bode plot is rela-

Figure 10.5

Circuit considerations that contribute to the control-to-output characteristics (forward-mode).

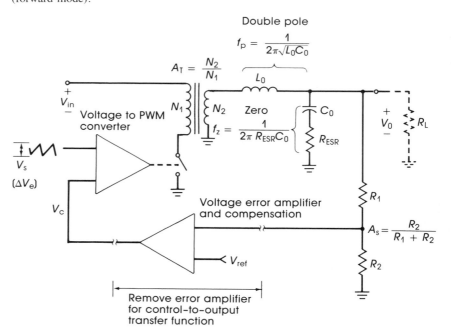

tively straightforward. The information needed to calculate the gain at DC and the pole and zero corner frequencies can be extracted from the schematic and the capacitor's data sheet.

10.4.1 Forward-Mode Control-To-Output Transfer Functions (Voltage-Mode Control)

The forward-mode converter, whether it is a buck or some form of transformer-isolated forward-mode converter using voltage-mode control, has a control-to-output transfer function similar to that shown in Figure 10.6. The lowest encountered corner frequency is that contributed by the output L–C filter. Its corner frequency can be calculated by

$$f_{LC} = \frac{1}{2\pi\sqrt{L_0 C_0}} \qquad (10.7)$$

This represents a double pole and exhibits a $-40-$dB/decade rolloff and a $-180°$ phase shift above its corner frequency. Since it can also be seen as an L–C tuned circuit, its Q has an influence on the Bode plot, although the Q does not generally affect the supply's performance. It can, however, exhibit a ringing response to a transient load condition if the phase is ever less than $30°$ below the closed-loop gain crossover frequency. The second major corner frequency is the zero contributed by the ESR of the output filter capacitor and output capacitance value itself. Its corner frequency is calculated by

$$f_z = \frac{1}{2\pi R_{ESR} C_0} \qquad (10.8)$$

This zero often is neglected by the designers and may present a hidden danger to the ultimate supply stability. The quality of output filter capacitor has a direct bearing on the placement of this zero's corner frequency. Poorer-quality capacitors have widely disparate ESR values, which can make the closed-loop response inconsistent between power supplies.

To calculate the gain of the control-to-output transfer function at 0 Hz (or DC) for a forward-mode converter, one simply adds the gain contributed by the step-up/step-down transformer to the gain of the time-averaged waveform presented to the L–C filter. This can be obtained by

$$A_{DC} = \frac{V_{out}}{V_{in}} = \frac{N_{sec}}{N_{pri}} \cdot \delta = \frac{V_{in}}{\Delta V_e} \cdot \frac{N_{sec}}{N_{pri}} \qquad (10.9)$$

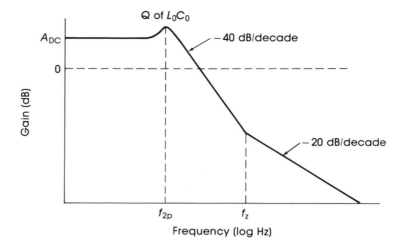

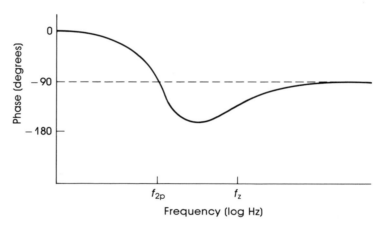

Figure 10.6

Control-to-output characteristics of forward-mode converters (voltage control).

V_e is the maximum change in voltage of the error amplifier to go from 0 to 100 percent duty cycle. This calculation should be done at the highest specified input voltage because the system exhibits its highest DC gain at that point. The Bode plot for the low input line case should also be drawn as a final check, but it usually does not present a problem. If a buck converter is used, the turns ratio for the transformer equals 1.

Now the gain and phase Bode plot for the forward-mode converter can be plotted with the real gains, corner frequencies, and phase changes of the system being designed. The compensation of the error amplifier can now be undertaken.

10.4.2 Flyback-Mode and Current-Mode Controlled Forward Converters

Control-to-Output Transfer Functions

Flyback-mode converters (see Fig. 10.7) using voltage-mode control and forward-mode converters using current-mode control have quite different control-to-output transfer functions than discussed previously. The main filter pole in the transfer function is heavily dependent on the equivalent resistance of the output load. In this case, it is recommended that the designer plot the Bode plot at high and low input line and at light and heavy loads.

The lowest corner frequency is a single pole created by the output

Figure 10.7

Circuit considerations that contribute to the control-to-output characteristics of flyback converters.

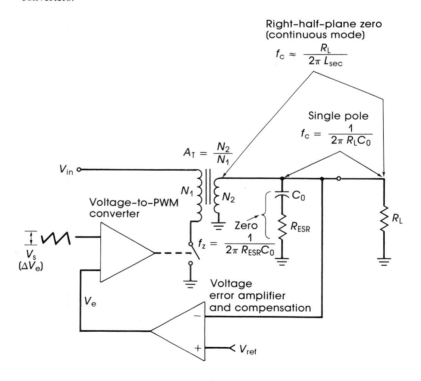

Right-half-plane zero
(continuous mode)

$$f_c \approx \frac{R_L}{2\pi L_{sec}}$$

Single pole

$$f_c = \frac{1}{2\pi R_L C_0}$$

$$A_T = \frac{N_2}{N_1}$$

V_{in}

N_1 N_2

Voltage-to-PWM converter

V_s
(ΔV_e)

Zero

$$f_z = \frac{1}{2\pi R_{ESR} C_0}$$

C_0

R_{ESR}

R_L

Voltage error amplifier and compensation

V_e

$+$ V_{ref}

filter capacitor and the output load resistance. It can be calculated at a given operating point by

$$f_p = \frac{1}{2\pi R_L C_0} \qquad (10.10)$$

Note that it is highly dependent on the output load resistance, so this calculation should be conducted at the lowest and highest specified load in order to find the range of break points that need to be considered. Also, if the load has inductive characteristics, it adds another zero to the transfer function, which can pose a serious threat to the system stability. For resistive loads, this pole contributes a -20-dB/decade rolloff above its corner frequency.

The zero contributed by the ESR of the output filter capacitor and the output filter capacitance is determined by Equation (10.8). This zero, in this case, causes the gain function to flatten out to a 0-dB/decade slope. This slope must be compensated for by the error amplifier by adding a pole to its compensation.

The DC gain of the transfer function is calculated by

$$A_{DC} = \frac{(V_{in} - V_{out})^2}{V_{in} \cdot \Delta V_e} \frac{N_{sec}}{N_{pri}} \qquad (10.11)$$

This should also be calculated for the highest and lowest input voltage in order to provide information for the range of Bode plots controlled by the input voltage and output load resistance.

The representative control-to-output transfer function of a discontinuous flyback-mode converter is shown in Figure 10.8. This curve would be applicable to the flyback, boost, and buck–buck topologies only if they are operating in the discontinuous mode of operation. This plot is also applicable to the forward-mode converter operating with a current-mode control method.

Flyback-mode converters can also enter into the continuous mode of operation, which means that the choke or transformer does not empty itself of flux during each cycle. The continuous mode of operation can be designed in or can be encountered in a discontinuous-mode converter at low line and at full load. Continuous-mode flyback converters can be extremely difficult, if not impossible, to compensate within the error amplifier in a voltage-mode control system. The reason for this is the appearance of a right-half-plane zero in the transfer function. The right-half-plane zero contributes a positive slope influence in the gain curve as does the left-half-plane zero but causes a lagging phase contribution.

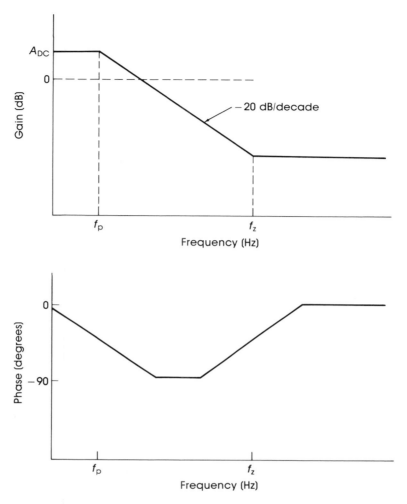

Figure 10.8

Control-to-output characteristics of a discontinuous flyback mode converter.

This is a real problem because it causes a rising gain slope and causes the phase to lag an additional $-90°$ at high frequencies. The typical error amplifier cannot provide enough phase lead to compensate for this severe phase lag. Current-mode control can help by improving the control-to-output phase performance by $+90°$, which makes the task of compensation a little easier. Its control-to-output transfer function is given in Figure 10.9.

The exact location of the right-half-plane zero is difficult to determine since it moves in frequency when the operating conditions change.

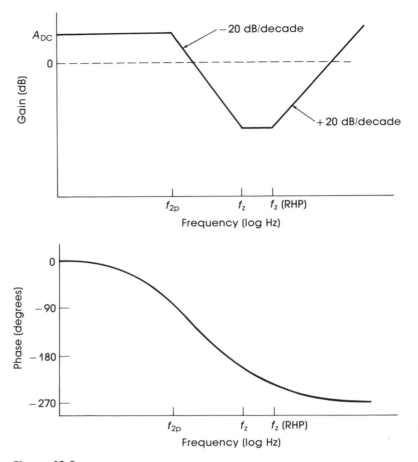

Figure 10.9

Control-to-output characteristics of continuous flyback-mode converters.

10.5 Common Error Amplifier Compensation Techniques

The purpose of adding compensation to the error amplifier is to counter-act some of the gains and phases contained in the control-to-output transfer function that could jeopardize the stability of the power supply. Obviously, the ultimate goal is to make the overall closed loop transfer function (control-to-output cascaded with the error amplifier) satisfy the stability criteria. This is to avoid having the closed-loop phase any closer to 360° than the desired phase margin anywhere where the gain is greater than 1 (0 dB). It is also desirable to have the slope of the gain curve at the crossover point with a value of −20 dB/decade. Phase margins of 45° to 60° (360° minus the total closed-loop phase lag) are considered safe values that yield well-damped transient load responses.

Three other considerations that affect the final power supply performance are:

1. The overall gain crossover frequency should be as high as possible in order to minimize the dynamic transient load response time.
2. The gain of the error amplifier at DC (0 Hz) should be as high as possible to ensure that the supply exhibits good load regulation.
3. A good practice is to make the average slope of the compensated closed-loop gain function close to -20 dB/decade.

Sometimes these considerations are at odds with the criteria for system stability, so expect some compromises on one or more of these considerations.

Real-world operational amplifiers exhibit a DC gain (dB) of between 80 and 110 dB. They also have a single pole with a corner frequency of 10 to 100 Hz and exhibit a gain rolloff of -20 dB/decade above the corner frequency. The compensation design cannot exceed this gain "envelope" presented by the operational amplifier. Usually the operational amplifier's frequency characteristics exceed the normal compensation requirements. It should, nonetheless, be checked.

10.5.1 Single-Pole Compensation

This type of compensation is used for converter topologies that exhibit a minimal phase shift prior to the anticipated gain crossover point. These include forward-mode regulators such as the buck, push–pull, and half- and full-bridge using either voltage or current-mode control techniques. These converters exhibit a relatively low phase shift below the pole contributed by the output filter. This compensation yields, though, a relatively poor transient load response time because the gain crossover frequency occurs at a low frequency. Its load regulation is very good, though, since its DC gain is very high. This method of error amplifier compensation is generally not used if a rapid transient load response time is desired.

The single-pole-compensated amplifier (Fig. 10.10) has a single pole at DC and rolls off at a -20-dB/decade rate forever. It also has a constant $-270°$ phase shift ($-180°$ by the inverting amplifier, and $-90°$ by the pole) over its entire frequency range. The error amplifier's gain crossover frequency is where the impedance of the input resistance equals the impedance of the feedback capacitor and is found by

$$f_{xo} = \frac{1}{2\pi RC} \qquad (10.12)$$

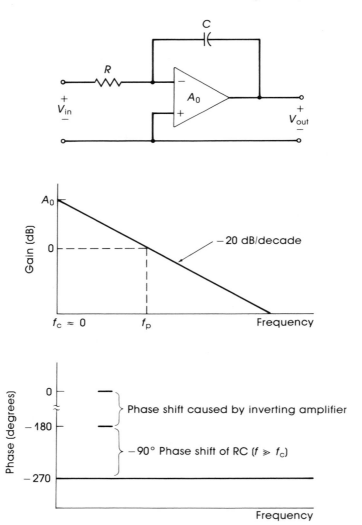

Figure 10.10
Single-pole compensation.

Its DC gain value is the open-loop gain of the operational amplifier since the feedback capacitor represents an infinite impedance at DC.

Figure 10.11 shows the Bode plot for a forward-mode converter using voltage-mode control and the single-pole compensation method.

The overall gain crossover point should be as close to the double pole of the output $L–C$ filter as possible and still exhibit a safe phase margin. Remember that when you select a desired phase margin, $45°$ to $60°$ is considered a reasonably stable range.

Since this form of error amplifier compensation has a constant $-270°$ phase lag across the entire frequency range, this allows only an addi-

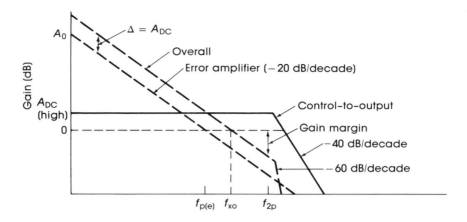

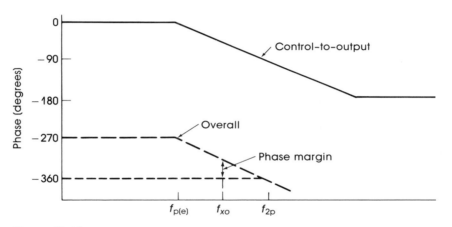

Figure 10.11

Plots showing compensation of a forward-mode regulator using a single-pole-compensated error amplifier.

tional 90° before the closed-loop system reaches 360°. This means that the double pole contributed by the L-C filter can contribute only 90° minus the desired phase margin. The only range of frequencies where this occurs is within the decade below the double L–C filter pole. By using a mathematical model that ignores the effects of Q on the phase plot one can calculate the worst-case phase margin;

$$\phi_m \geq 2 \, \tan^{-1}\!\left(\frac{f_{xo}}{f_p}\right) \qquad \text{(forward-mode regulators)} \qquad (10.13)$$

or

$$f_{xo} = f_p \, \tan\!\left(\frac{\phi_m}{2}\right) \qquad (10.14)$$

where f_p is the L–C corner frequency and f_{xo} is the overall closed-loop gain crossover frequency.

Solving for some representative safe phase margins, the resulting overall gain crossover frequencies become

$$60°: f_{xo} = 0.57f_p$$
$$45°: f_{xo} = 0.41f_p$$

This is the approximate maximum overall gain crossover frequency of the entire closed-loop system. The actual phase margin will be higher than the selected desired phase margin because of the effects of the Q of the L–C filter. It is not recommended to push the gain crossover frequency above this point, since the Q is very dependent on the ESR of the capacitor, which can vary greatly between manufacturers.

Now to calculate the error amplifier gain crossover frequency, the influence of the inherent DC gain of the control-to-output curve must be subtracted from the overall gain curve. This is done by

$$f_{xe} = f_{xo} \cdot 10^{-(A_{DC}/20)} \tag{10.15}$$

where f_{xe} is the error amplifier's gain crossover frequency. This value for f_{xe} can now be used in Equation (10.12) to determine the value of the feedback capacitor. As one can see, the error amplifier's gain crossover frequency is quite low, which will result in very slow transient load response time. For this reason alone it is not generally used by designers.

10.5.2 Zero-Pole Pair Compensation

This compensation method is used for converters that exhibit a single filter pole at low frequency and a maximum phase shift of 90°. These converters are the boost, buck–boost, and the flyback topologies if they are operating in the discontinuous mode of operation. Forward-mode converters with current-mode control are also included. The pole caused by the output filter capacitor and the load resistance occurs at an extremely low frequency. So if a single pole compensation method were to be used, the resulting loop bandwith would be less than 50 Hz, which is totally unacceptable. By introducing a zero below this first pole, the loop bandwith can be extended far beyond this corner frequency, thus vastly improving the overall transient load response time.

The zero-pole pair schematic and Bode plot are shown in Figure 10.12. It has a pole at DC as in the single-pole amplifier, but now has a zero, which flattens out the amplifier gain curve until an additional high-

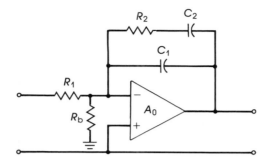

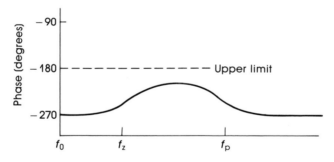

Figure 10.12

Zero-pole pair compensation.

frequency pole is encountered. There is also a corresponding "bump" in the phase curve that occurs during the flat gain portion of the gain curve. The peak of this phase bump should be located at the point of greatest phase lag in the control-to-output function, which occurs just above the output filter pole. The amount of phase lead is proportional to the width of the constant gain portion of the gain curve and has a maximum possible phase angle of $-180°$ (or $+90°$ bump).

The corner frequencies above DC are

$$f_z = \frac{1}{2\pi R_2 C_2} \tag{10.16}$$

$$f_p \cong \frac{1}{2\pi R_2 C_1} \tag{10.17}$$

The gain of the error amplifier at the flat gain section of the curve is

$$A_k = \frac{R_2}{R_1} \tag{10.18}$$

Now, how does one select the gain crossover point and the location of the pole and zero? The general purpose of introducing zeros and poles in the error amplifier is to counteract poles and zeros in the control-to-output function, respectively. This will yield an average -20-dB/decade closed-loop gain response (see Fig. 10.13) with the phase never getting to within the desired phase margin when the gain is greater than one. When one is selecting the gain crossover frequency, one must keep in mind that if the gain crossover frequency gets too close to the switching frequency, the amplifier begins to amplify too much of the ripple voltage. A safe value of crossover frequency is 20 percent of the switching frequency:

$$f_{xo} = \frac{f_s}{5} \tag{10.19}$$

where f_s is the frequency of operation of the switching power supply.

Next, the designer must determine the gain needed by the amplifier to bring the control-to-output curve up to the zero decibel level at the desired crossover frequency. This is simply done by seeing the amount of attenuation exhibited by the control-to-output curve at the desired crossover frequency.

$$A_{xo} = \frac{1}{H(s)}\bigg|_{f = f_{xo}} \tag{10.20}$$

Next, determine the worst-case phase lag of the control-to-output function below the overall gain crossover frequency. The error amplifier must not only overcome this phase lag but also "add in" additional phase lead to provide for the desired closed-loop phase margin. The minimum desired "amplitude" of the phase bump can be calculated by

$$\phi_{bump} = 2 \tan^{-1}\left(\frac{\sqrt{f_p}}{\sqrt{f_z}}\right) - 90° \tag{10.21}$$

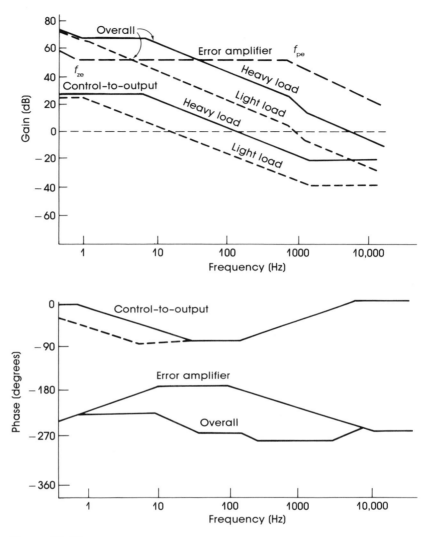

Figure 10.13

Zero-pole pair compensation of a discontinuous flyback-mode regulator.

The zero is placed at the same frequency of the pole caused by the output filter capacitor and load resistance. In flyback converters, this pole shifts in frequency proportional to the load resistance. The lowest frequency is found at the highest load resistance (or lightest load), so this is where the zero should be placed. If this is done, the $-45°$ phase lag in the control-to-output function is counteracted by the $+45°$ phase lead of the error amplifier.

The high-frequency pole is placed to counteract the zero caused by the ESR of the output capacitor and capacitance value itself. The pole should be placed at what would be the lowest anticipated ESR zero frequency. This would be the worst ESR value for the type of capacitor and manufacturer chosen.

Once the zero and pole frequencies are determined, one can calculate the values of the compensation directly from the Bode plot. The input resistor (R_1) is known because it is the upper resistor in the voltage-sensing resistor divider. The lower resistor does not enter into the compensation scheme but is considered only when minimizing the effects of the input bias current of the operational amplifier. The value of the feed-back components can be found as follows.

$$C_1 = \frac{1}{2\pi f_{pe} A'_{xo} \cdot R_1} \tag{10.22}$$

$$R_2 = A'_{xo} \cdot R_1 \tag{10.23}$$

$$C_2 = \frac{1}{2\pi f_{ze} \cdot R_2} \tag{10.24}$$

where f_{ze} is the frequency of the zero, f_{pe} the frequency of the high-frequency pole, and A'_{xo} the absolute gain needed at the overall crossover frequency. The boost exhibited by the amplifier at the center of the constant-gain portion of the gain curve is

$$\text{Boost} = 2 \tan^{-1}\left(\frac{\sqrt{f_{pe}}}{\sqrt{f_{ze}}}\right) - 90° \tag{10.25}$$

The Bode plot (see Fig. 10.13) for the overall closed-loop response should be drawn to verify the gain slope at the crossover frequency and the variation in crossover frequency with various line, loads, and ESR values. If the gain slope is -40 dB/decade at any of the operating conditions at the crossover frequency or if the phase margin is less than the desired result, some adjustments to the upper pole location may have to be undertaken. Typically, though, these component values will be satisfactory.

10.5.3 Two-Pole–Two-Zero Compensation

This method of compensation is intended for power supply converters that exhibit a -40-dB/decade rolloff above the poles of the output filter and a $-180°$ phase lag. These are the forward-mode converters such as the buck, push–pull, half-bridge, and full-bridge topologies using

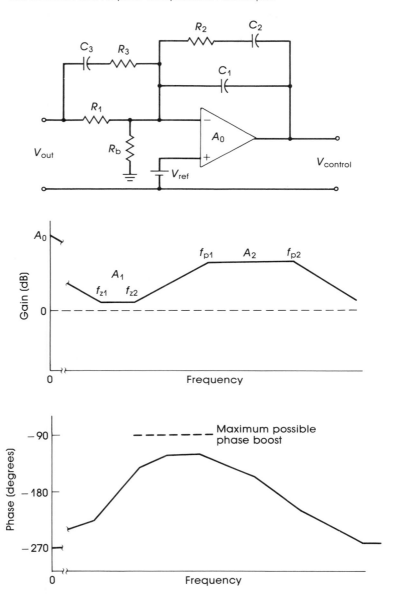

Figure 10.14

Two-pole–two-zero error amplifier compensation.

voltage-mode control. Like the zero-pole pair compensation method, the two-zero–two-pole method introduces zeros into the error amplifier compensation to reduce the steep gain slope above the double pole caused by the $L-C$ filter and its associated $-180°$ phase shift.

The schematic and Bode response are given in Figure 10.14. As one can see, the amplifier has a +20-dB/decade slope between the zero and pole pairs. It also has a "phase bump" that peaks at the geometric mean frequency between the highest pole and the lowest zero frequency. The phase bump also has a maximum possible phase lead of +180° (or a net amplifier phase lag of −90°). The actual peak value of the phase bump is determined by the frequency separation between the zeros and the poles. So the greater the difference, the higher the phase peak. The purpose of the phase peak is to counteract the −180° phase lag of the L–C filter. The +20-dB/decade slope over this range also brings the overall gain slope to the −20-dB/decade slope. The lower of the high-frequency poles within the amplifier counteracts the zero caused by the ESR of the filter capacitor. The last pole is placed to increase the gain margin of the closed-loop system. So this pole is placed at a frequency higher than the gain crossover frequency. (See also Fig. 10.15.)

The points of interest within this compensation method are (A' is the absolute gain)

$$A'_1 = \frac{R_2}{R_1} \tag{10.26}$$

$$A'_2 = \frac{R_2(R_1 + R_3)}{R_1 R_3} \approx \frac{R_2}{R_3} \tag{10.27}$$

$$f_{z1} = \frac{1}{2\pi R_2 C_2} \tag{10.28}$$

$$f_{z2} = \frac{1}{2\pi(R_1 + R_3)C_3} \approx \frac{1}{2\pi R_1 C_3} \tag{10.29}$$

$$f_{p1} = \frac{1}{2\pi R_3 C_3} \tag{10.30}$$

$$f_{p2} = \frac{C_1 + C_2}{2\pi R_2 C_1 C_2} \approx \frac{1}{2\pi R_2 C_1} \tag{10.31}$$

$$\text{Boost} = 4 \tan^{-1}\left(\sqrt{\frac{\sqrt{f_{p1}}}{\sqrt{f_{z2}}}}\right) - 180° \tag{10.32}$$

Once the corner frequencies of the control-to-output transfer function have been determined, the gain and the placement of the zeros and poles of the compensation can be determined. This is easily done as follows.

1. Determine the crossover frequency by

$$f_{xo} = \frac{f_s}{5} \quad \text{(highest)} \tag{10.33}$$

where f_s is the supply's switching frequency.

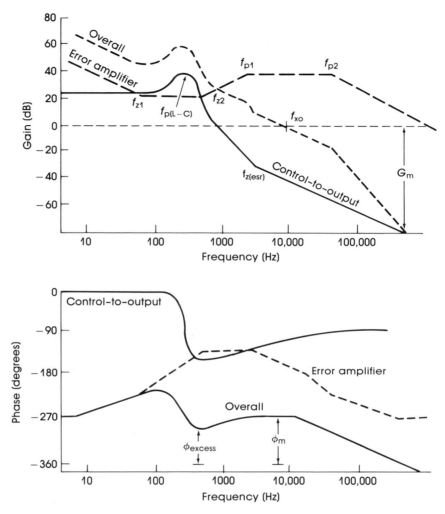

Figure 10.15
Separating the pairs in the two-pole–two-zero compensation method.

2. Determine the gain needed to bring the control-to-output transfer function up to 0 dB at the desired crossover frequency.
3. If both zeros are placed at the same frequency, then they should be placed at

$$f_{z1} = f_{z2} = \frac{f_{p(L-C)}}{2} \tag{10.34}$$

The resulting margin in excess phase is 45°. If a greater value is desired, move the zero pair lower in frequency. Or to deemphasize the Q of the L–C filter on the overall gain function, the lower zeros can be separated. In this case

$$f_{z1} = \frac{f_{p(L-C)}}{5} \tag{10.35}$$

The second zero is placed just above the L–C poles such that

$$f_{p(L-C)} < f_{z2} < 1.2 f_{p(L-C)} \tag{10.36}$$

This reduces the amplitude peak caused by the Q of the output filter by not providing additional gain over this region.

4. Place the f_{p1} at the lowest expected zero caused by the ESR. This will be caused by the highest expected ESR value.

$$f_{p1} = f_{z(ESR)[\text{worst case}]} \tag{10.37}$$

5. Place f_{p2} above the overall crossover frequency at approximately

$$f_{p2} > 1.5 f_{xo} \tag{10.38}$$

Now the component values of the compensation network can be determined. Resistance R_1 is known, since it is the upper resistor in the resistor divider of the output-voltage-sensing network. The gain at the first zero (f_{z1}) can be calculated by

$$A_1 = A_{xo} + 20 \log \left(\frac{f_{z2}}{f_{p1}} \right) \qquad [\text{in dB}] \tag{10.39}$$

Other component values are determined as ($A'_\square$ is the absolute gain)

$$C_1 = \frac{1}{2\pi f_{xo} A_1 R_1} \tag{10.40}$$

$$R_2 = A'_1 \cdot R_1 \tag{10.41}$$

$$C_3 = \frac{1}{2\pi f_{z2} R_1} \tag{10.42}$$

$$R_3 = \frac{R_2}{A'_2} \tag{10.43}$$

$$C_2 = \frac{1}{2\pi f_{z1} R_2} \tag{10.44}$$

The worst-case excess phase occurs just above the L–C filter double pole, which will be greater than 45°. The phase margin at the crossover frequency will be 90°. If an oscillation problem were to occur, it would occur at the resonance of the L–C filter, but 45° and the lowered gain in this region are sufficient to ensure that this will not happen.

If the procedure outlined above is followed, then system stability performance will usually be satisfactory. Minor adjustments may be necessary.

10.6 Attempting to Compensate for a Right-Half-Plane Zero

Continuous-mode, flyback-mode switching power supplies exhibit an additional difficulty in trying to stabilize the system, and this is caused by a right-half-plane zero. A right-half-plane zero causes a +20-dB/ decade gain influence just like the left-half-plane zeros, previously discussed, but exhibits a −90° phase shift. This is extremely difficult to compensate. If one attempts to compensate for the rising gain, the phase reaches 360° and the supply oscillates. If one compensates for the phase, then the overall gain increases with frequency and the gain crossover frequency becomes difficult to attain. The only easy alternative is to lower the gain crossover frequency far below the desired value. This makes the continuous-mode, flyback-mode converters virtually impossible to compensate using any control method.

The physical manifestation of a right-half-plane zero can be visualized when one looks at the combined input and output transformer current waveforms, that is, if the transformer were a 1:1 turns ratio or a simple boost inductor as in Figure 10.16. When a flyback-mode regu-

Figure 10.16

Physical manifestation of the right-half-plane zero in flyback mode converters.

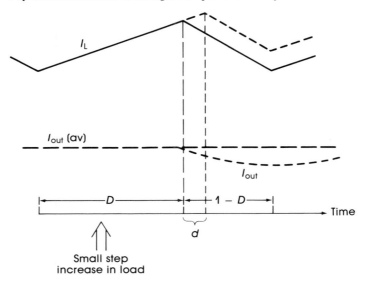

lator operates in the continuous mode, the transformer or inductor core does not empty itself of flux energy. This shows up as a current pedestal on top of which the current ramp rides. The load current is the rectifier's current waveform averaged over the entire operating period. If a step increase in load occurs, the error amplifier causes the converter to increase its "on-time" duty cycle. During the first few cycles, the peak current has not had time to increase to the higher value necessary to provide the increased DC load current. What actually happens is that the lowered on-time of the rectifier causes the output current to drop instead of increase. This continues until the peak current increases to accommodate the new load requirement.

The location of the right-half-plane zero moves in frequency when the operating conditions change. This makes compensation difficult. The location of this zero at any one operating condition can be found by

$$f_z(\text{RHP}) = \frac{R_L V_{\text{in}}^2}{L_{\text{sec}} V_{\text{out}}(V_{\text{in}} + V_{\text{out}})} \tag{10.45}$$

where R_L is the equivalent load resistance ($V_{\text{out}}/I_{\text{out}}$), V_{out} the output voltage (DC), V_{in} the input voltage (DC), and L_{sec} the inductance of the secondary.

It is recommended that the gain crossover frequency be placed well below the worst-case lowest frequency of the right-half-plane zero. A left-half-plane pole should be used to counteract the rising gain characteristic of the right-half-plane zero when it is above the crossover frequency. This will prevent the gain from reemerging above 0 dB. The phase will, of course, cross 360°, but the gain is guaranteed to be less than 0 dB. Most designers simply do not use the continuous-mode, flyback-mode converters, and for good reason.

11

Resonant Converters—
An Introduction

During 1972–1979 the PWM-type switching power supply had remained at a relative plateau in regard to their advances in performance. Their operating frequencies were remaining well below 100 kHz and their power densities (watts per cubic inch) remained relatively constant. The factor that slowed the PWM switching power supply in its evolutionary path was its internal losses that were dictated by the state of the technology at that period. During the late 1970s, the development of the first practical, commercially priced power MOSFETs provided a step improvement in efficiency and power density. All at once, the switching losses dropped significantly. This allowed the PWM switching power supplies to reach 100 kHz before efficiencies began once again to drop off. Once again the switching loss and core losses (hysteresis and eddy-current losses) now became a significant portion of the total losses. Meanwhile, in the other electronics fields, the circuitry continued its march toward miniaturization. For the switching power supply field to keep pace with the rest of the fields, something would have to be done other than just simply making improvements in the components. This began to occur 1981. Serious research had begun in defining and characterizing resonant modifications to the basic PWM switching power supply topologies. Its ultimate goal was to drastically reduce the switching losses within the switching power supply, which is the dominant frequency-dependent loss within the supply. With switching losses virtually eliminated, the core losses would be the only major obstacle in raising the switching power supply's operating frequency to the point where a significant reduction in size could take place.

Another development that occurred in this period that would help both the PWM and resonant supplies took place in component packaging technology. Surface-mount technology was placing the low-level control circuitry in packages about one-fifth their former volume. Their superior high-frequency characteristics and compact layout density permitted the control section of the power supply to significantly reduced in size.

Today, both the PWM and the resonant fields are benefiting from the research that has taken place within the resonance field. Some commercially available PWM switching power supplies have now reached operating frequencies of 500 to 800 kHz. To accomplish this, the PWM designers had to pay close attention to the RF considerations of the design and truly understand the components they were using.

Practical, commercially available resonant power supplies are emerging in the marketplace today, but the field of resonant structures is still very much in a state of change. New demands placed on the designers and component manufacturers at these higher frequencies are resulting in improved modeling, better high-frequency components, and better high-frequency materials and packages, all with the ultimate goal of improving efficiency to further reduce the size of the power supply.

11.1 Why Resonant Switching Power Supplies?

The factor that limits the ultimate size of a switching power supply is its efficiency. Efficiency is the amount of power the supply must use (or waste) in order to provide power to its loads. The major portion of this internally generated power is converted into heat by the various components that produce $E \times I$ products across their terminals. The major heat-producing losses within a PWM switching power supply are switching and conduction losses within the semiconductors, core losses, and resistive losses distributed over the entire supply (see Fig. 11.1). The major task of the designer is how to effectively rid the supply of this heat and dispose of it to the outside world. So the designer faces the trade-off of how much heat can be generated, which affects the operating frequency and the component size, versus how effective is the environment in accepting the heat given to it.

This is where the resonance technology makes a quantum leap in the designer's favor. It drastically reduces the switching losses within the supply, which is one of the top heat-generating losses. This removes 30 to 40 percent of the losses within a comparable PWM supply when operated at the same frequency (see Figs. 11.2 and 11.3). The designer can then increase the operating frequency in order to reduce the major component sizes, hence increasing the power density. As a result,

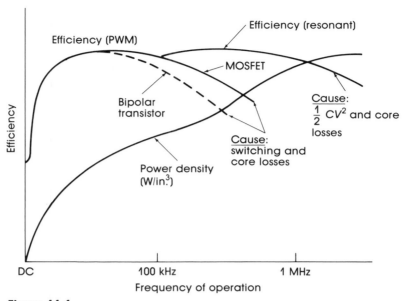

Figure 11.1
Power supply trends over frequency.

switching power supplies with operating frequencies of 500 kHz to 15 MHz are now possible.

An added advantage that is in the designer's favor is the significant reduction in RFI/EMI. By eliminating the very rapid transitions in current and voltage, the harmonic-rich waveforms are also eliminated. This makes it easier to pass the RFI requirements imposed by the approval bodies with less RFI filtering. A sigh of relief for those designers who live in terror during the design and preproduction phases of a development program.

The disadvantages are that the resonant converters are more complex than their PWM counterparts, and consequently require a longer time to design and cost more to implement. The IC manufacturers have been waiting for the resonant supply field to settle down to see which form of control topology would be most applicable to the designers. So highly integrated control ICs are just now emerging in the marketplace. If none of these ICs fit your needs, the control function may have to be built from discrete components.

Acceptance of the resonant supply topologies has been slow. Many companies have taken a "wait and see" position until more definitive structures emerge from the field. Nonetheless, resonant-type supplies will eventually offer the key to miniaturizing power supplies.

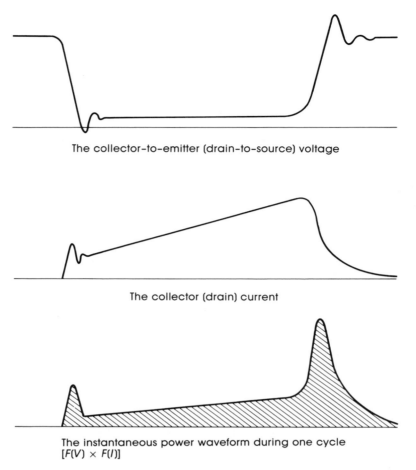

The collector-to-emitter (drain-to-source) voltage

The collector (drain) current

The instantaneous power waveform during one cycle
[$F(V) \times F(I)$]

Figure 11.2
Power losses within a conventional PWM power switch.

11.2 Basic Quasi-Resonant Converter Operation

Resonant converters, as the general field is called, are better described by the term "quasi-resonant" converters. In the pure sense of the word, resonance is a continuous function whose waveform is a continuous sinusoid signal, and, indeed, there are topologies that utilize the full (or continuous) resonance technique. This does not appropriately describe the operation of most present-day resonant converters that are appearing on the market today. The term "quasi-resonance" or "discontinuous resonance" better indicates their mode of operation. Quasi-resonant converters do exhibit a resonance in their power section, but instead of

The collector-to-emitter (drain-to-source) voltage

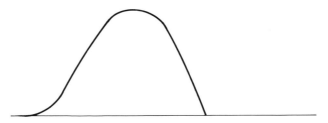

The collector (drain) current

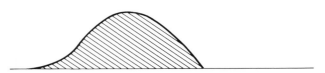

The instantaneous power waveform during one cycle
$[F(V) \times F(I)]$

Figure 11.3
Power losses within a quasi-resonant switch (half-wave).

the resonant elements being operated in a continuous fashion, they are operated for only one-half of a resonant sine wave at a time. Its operation is based on the step response of a resonant, $L-C$ tank circuit that "rings" at its resonant frequency described by

$$f_r = \frac{1}{2\pi\sqrt{LC}} \qquad (11.1)$$

The power switch in quasi-resonant converters connects the input voltage source to the tank circuit, and is turned on and off in the same step fashion as in PWM switching power supplies. The conduction period (or "on" period) is highly dictated by the ringing frequency of the tank

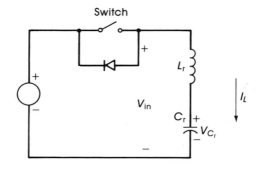

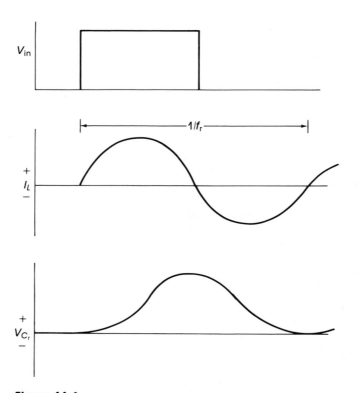

Figure 11.4

Pulse response of a series tank circuit.

circuit (see Fig. 11.4). The power switch turns off after the completion of one-half of a resonant period (one-half cycle of a sine wave). What this accomplishes is that the current at the turn-on and turn-off transitions is zero, thus eliminating the switching loss within the switch.

Now let's examine how one applies this phenomenon to convert power from its input to its load. Any load placed on the tank circuit should not be heavy since it will reduce the ability of the tank circuit to ring by reducing its Q. If a circuit is placed in parallel with any of the resonant elements, it should represent a high impedance such as the input of an L–C lowpass filter. Let's examine the case when one places the L–C filter in parallel with the resonant capacitor (Fig. 11.5). If the cutoff frequency of the L–C filter is much lower than the resonant frequency of the tank circuit, the filter will time-average the resonant capacitor's voltage as it does in a PWM forward-mode, buck regulator.

Figure 11.5
Responses of a buck-loaded series tank circuit.

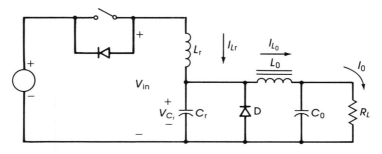

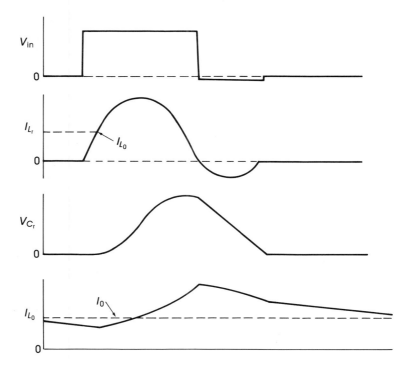

The tank circuit always starts from an initial condition, between power switch conduction cycles, of no current flowing through the resonant inductor and no voltage across the resonant capacitor. When the power switch turns on, the current through the resonant inductor cannot change instantaneously and remains at zero current throughout the turn-on transition. Because the resonant capacitor is effectively clamped by the forward-biased commutation diode, the resonant inductor's current begins to ramp up with a slope of V_{in}/L_r until the resonant inductor current exceeds the load current flowing through the commutating diode, at which point the diode turns off. Now the low impedance of the forward-biased commutation diode is replaced by the high input impedance of the $L–C$ filter and the tank circuit is permitted to "ring" in a true sinusoidal fashion. The resonant capacitor's voltage lags the resonant inductor's current by 90°, so the inductor's current returns to zero as the resonant capacitor's voltage is at its peak. The power switch then turns off. The input of the $L–C$ filter resembles a constant-current sink, which then removes the remaining charge from the resonant capacitor, resulting in a negative linear voltage ramp. When the resonant capacitor's voltage reaches zero, the commutating diode turns on in order to maintain the load current through the $L–C$ filter. While this is happening, the current in the resonant inductor tends to flow backward through the power switch. It is permitted to do so by the existence of the antiparallel diode across the power switch. This excess energy not removed by the output $L–C$ filter is returned to the input bulk filter capacitor and recovered for future use. The tank circuit now returns to its initial condition, ready for the next resonant conduction cycle.

As one can see, the product of voltage and current waveforms at the power switch's terminal is virtually zero during the switching transitions. The commutating diode similarly has zero current flowing through it during its switching periods, but in a different fashion. At turn-on, the resonant inductor current ramps up, displacing the load current through the commutating diode until all the load current can be sustained by the tank circuit. The diode, at that instant, has no current flowing through it. At turn-on, the diode exhibits a negligible loss due to the forward recovery time of the diode, but it is less than in a comparable PWM buck converter. This is the basic mode of operation for a zero-current, forward-mode, quasi-resonant converter. There are variations in the arrangement of the components because of their topological requirements, but the operation is basically analogous.

Since the "on" period of the power switch must be fixed to the resonance period of the tank elements, the regulation of the output power is accomplished by varying the repetition rate of the on-times. This is normally referred to as *fixed on-time, variable off-time control,* but in the resonance field it is referred to simply as *variable-frequency control.* Within this mode of control, the output power is controlled by the number of "on" cycles per second, so to increase the output power, the controller increases the frequency. This type of control yields a linear control relationship.

$$\frac{V_0}{V_{in}} \cong \frac{f_s}{f_r} \qquad (11.2)$$

But this type of control does have its limitations. A low-frequency limit must be imposed on the control network to avoid dropping close to the pole of the output filter. This would result in excess output ripple voltage at the pulse repetition frequency (PRF) of the supply. A high-frequency limit must also be imposed to prevent the power supply from entering the continuous mode of operation. At that point, the power switch and rectifiers are forced to switch before the resonant elements can be emptied of their energies. This reintroduces the switching losses into the semiconductors, which defeats the purpose of quasi-resonant switching.

The need for a precise "on" period for the power switch is relaxed by the use of the antiparallel diode. The diode is already resident within the MOSFET itself and may be used for this purpose, if the diode has a good forward voltage drop characteristic. The diode plays an important role within the supply: specifically, to shunt the "ring-back" current away from the power switch prior to its turn-off. This allows the power switch to not only turn off at any time during the "ring-back" period but also have very slow turn-off characteristics. The demands placed on the power switch are even less stringent than in PWM switching power supplies.

As one can see, the operation of the basic quasi-resonant switching power supply is analogous to the operation of the PWM buck regulator except the switching waveform has been preshaped to promote the elimination of the semiconductor switching losses. Schematically, and electrically, the quasi-resonant converter is relatively easy to understand. The difficulties arise from knowing which topologies can offer an optimum solution to the requirements, knowing what reasonable assumptions can be made during the design process, and understanding the high-frequency behavior of the elements within the design. As I had alluded to earlier, the field of resonant switching power supplies is still quite new. New research is experimenting with varying the topologies to minimize the high-frequency parasitic effects and to further reduce the size. The result is a sea of research papers with few definitive design rules to help the everyday designer implement a producible resonant power supply. Hopefully, the following text will provide an intuitive foundation for the reader to understand the ongoing evolution within the field.

11.3 The Resonant Switch—A Method of Creating a Quasi-Resonant Family

The topologies within the quasi-resonant converter field are simply resonant elements added to many of the basic PWM topologies. There are forward-mode converters, such as the buck, boost, half-forward, and half-bridge, and there are flyback-mode converters. The transformer and filter inductor within the resonant supplies are not used as the reactive elements responsible for the resonance phenomenon. Instead, additional reactive elements are added in order to produce the resonant current and voltage waveforms. The transformer and filter inductor are used for stepping up or down and filtering just as they were in the PWM family of supplies. To convert a PWM supply to a quasi-resonant supply one replaces the power switch (schematically) with a resonant switch (Lee and Liu, 1986), as well as replacing the control circuit. The resonant switch is simply a power switch placed within a resonant network as in figure 11.6. The resonant switch exhibits the ability to output a sinusoidal voltage whenever the power switch is turned on.

As one can see, the arrangement of the zero-current resonant switch can vary slightly. Both arrangements have the same equivalent AC models. The resonant capacitor returns to a low-impedance AC ground, which either side of the input voltage source represents. The output of the zero-current, quasi-resonant switch is taken from the terminal voltage across the resonant capacitor.

By using the resonant switch technique, one can create a family of zero-current, quasi-resonant switching power supplies. Figure 11.7 shows some of the zero-current, quasi-resonant family members that can be created from topologies within the PWM converter family. In fact, all the PWM topologies can be converted in this fashion. The push–pull topology is the only one that is difficult to convert to a quasi-resonant topology because of the center-tapped arrangement of the primary wind-

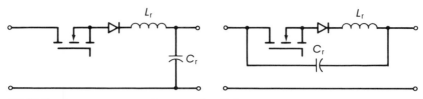

(A) Half-wave, zero-current resonant switches

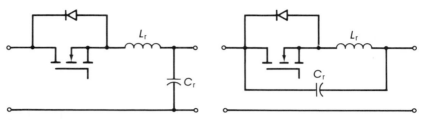

(B) Full-wave, zero-current resonant switches

Figure 11.6
Zero-current, resonant switches.

ing. The push–pull topology would require two resonant switches, which would cost more and be difficult to control since the tolerances on the resonant elements would cause their resonant periods to differ. As one can see in Figure 11.7, the modification to the basic PWM topology is quite minor.

To keep the current within the power switch at zero during its turn-off, the diode was added to the power switch. The parallel arrangement discussed previously allows the excess energy within the resonant elements to be returned to the input capacitor through the diode. This arrangement is called a *full-wave resonant switch* (Lee and Liu, 1986). An entire period in the resonant current is permitted to flow through the power switch–diode circuit. Another arrangement that would prevent current from flowing through the power switch during the ring-back period would be to place a diode in series with the power switch. This arrangement is called a *half-wave quasi-resonant* switch (Fig. 11.6) The diode now blocks the ring-back current. The power switch would also have zero current flowing through it during its turn-off, but now the excess energy in the resonant elements exists without a convenient place to go. The energy will eventually enter the output filter, but the inductor is unloaded and may radiate some of this energy to the environment in the form of RFI. The full-wave resonant switch is more commonly used by resonance designers.

The resonant frequency of the quasi-resonant switch should be well

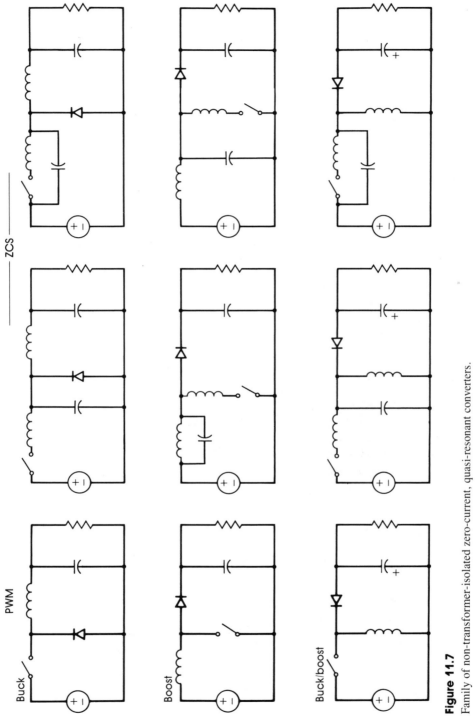

Figure 11.7

Family of non-transformer-isolated zero-current, quasi-resonant converters.

above the poles of the output filter. The resosnant switches within to-day's commercially marketed quasi-resonant power supplies have frequencies falling in the range of 500 kHz to 1 MHz. This is driven mainly by the technology of the control circuitry, which must maintain the 500-nsec pulsewidth for the power switch. Technologies are presently being developed that will allow the resonant switches to operate to the 10- to 20-MHz region with less fear of tolerance buildup and temperature drift.

The quasi-resonant switch provides a convenient lumped modeling technique in the design of quasi-resonant converters.

11.4 The Zero-Voltage Quasi-Resonant Converter Family

The ultimate purpose of resonant switching power supplies is to eliminate the switching losses within the semiconductors. This means that the voltage–amperage product during the switching transitions is equal to zero. The zero-current, quasi-resonant switch allows step transitions in its terminal voltage when there is no current flowing through itself. The same criteria of a zero voltage–amperage product could be accomplished by making the voltage zero during the switching transitions. Using the quasi-resonant switch technique, a whole new family of zero-voltage, quasi-resonant converter topologies can be created.

The zero-voltage quasi-resonant switch now takes the form of that diagrammed in Figure 11.8. Note that the resonant capacitor has been moved from the output of the resonant switch to the terminals of the power switch. Its operation is completely opposite to that of the zero-current, quasi-resonant switch. Instead of the current through the inductor starting from an initial zero condition, with its current lacking the ability to change instantaneously, the capacitor starts at an initial voltage, which places zero volts across the power switch at turn-on, and its voltage cannot change instantaneously. Its method of control is also completely opposite. The zero-voltage quasi-resonant switch is operated in a fixed off-time, variable on-time method of switching. So for a resonant switching cycle, the zero-voltage quasi-resonant switch begins with the power switch in the "on" condition and then turns off to begin the switching cycle. Referring to Figure 11.9, we see that the voltage across the resonant capacitor proceeds negatively through a sinusoidal waveform until it returns to $+V_{in}$. At the point where the sinusoidal voltage should exceed the level of the input voltage, the ring-back diode, anti-parallel to the power switch, begins to conduct and the excess energy is

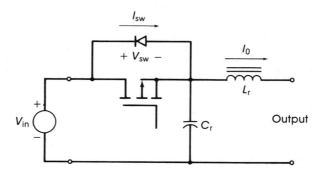

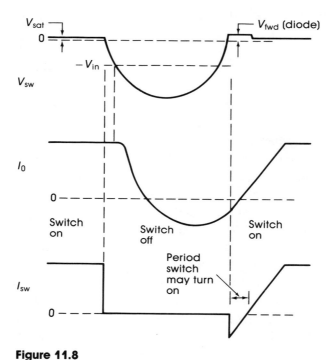

Figure 11.8
The zero-voltage, quasi-resonant switch and waveforms.

returned to the input and recovered. During that ring-back period, the power switch may turn on again and accomplish zero-voltage switching. Immediately following the power switch turn-on, the current through the series resonant inductor quickly proceeds through a linear ramp as dictated by

$$I_{sw(1)} = \frac{V_{in}}{L_r} \qquad (11.3)$$

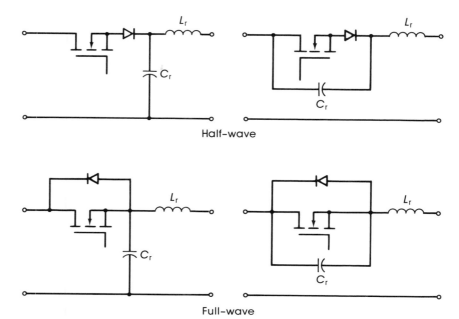

Figure 11.9
Zero-voltage, quasi-resonant switches.

As seen in Figure 11.10, this linear current ramp continues until its current exceeds the load current that was being sustained by the commutating diode. At that point the diode turns off (at zero current) and disappears from the circuit. Now the resonant inductor is in series with the filter inductor, and its value may be added to the value of the output filter inductor. Since the resonant capacitor now has V_{in} across it and the inductor has been assimilated by the output filter inductor, the entire zero-voltage, quasi-resonant switch has disappeared from the power supply. The slope of the linear current ramp then drops to a slope dictated by the output filter inductor:

$$I_{sw(2)} = \frac{V_{in} - V_{out}}{L_o + L_r} \tag{11.4}$$

This condition continues until the next resonant cycle is begun by turning off the power switch.

The zero-voltage, quasi-resonant converter uses a constant off-time, variable on-time method of control. The control "sense" is completely opposite to that of the zero-current converter. For heavier loads, the number of resonant off-periods per second is low. For light loads, the number of resonant off-periods per second will be high but cannot

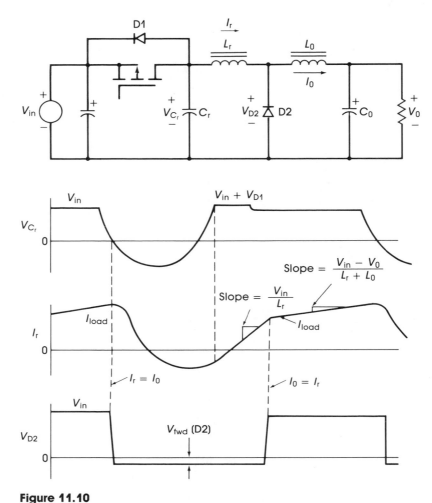

Figure 11.10

Waveforms within a zero-voltage, quasi-resonant buck converter.

go any higher than the resonant frequency of the tank circuit. So the zero-voltage converter must have a minimum load on its output to maintain the zero-voltage switching criteria and the integrity of the control loop. Theoretically, the zero-voltage converter has an unlimited overload capability where the control frequency can drop to zero, but practically there must be an overcurrent limit that keeps the control frequency above a minimum frequency. This low-frequency limit is dictated by the DC forward-biased safe operating area (FBSOA) of the power switch and the thermal limits of the windings in the series inductors.

By utilizing the quasi-resonant switch technique, one can create an entire family of zero-voltage, quasi-resonant converters. Figure 11.11

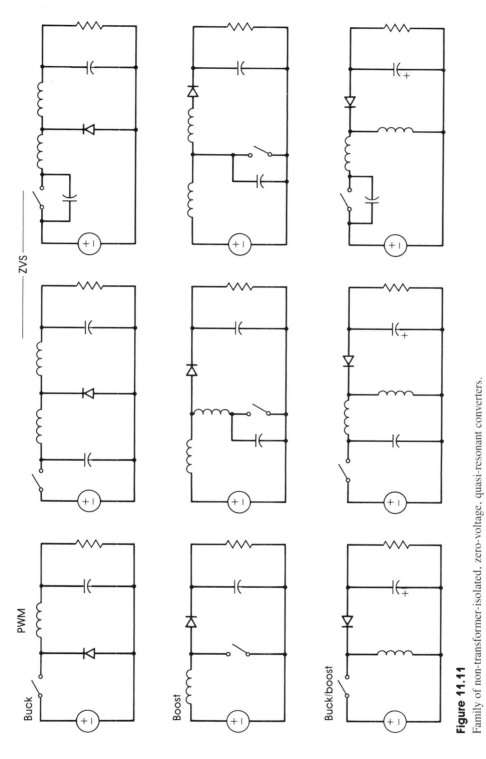

Figure 11.11

Family of non-transformer-isolated, zero-voltage, quasi-resonant converters.

shows some of the non-transformer-isolated topologies that can be created by this method. Transformer-isolated family members can be added to these, but they will be discussed later.

The zero-voltage converters can be viewed as inverse duals of the zero-current converters. The current and voltage waveforms are switched between the two families, and the control method is also completely opposite. One family may offer benefits over the other in certain applications, so they should be closely considered when choosing a resonant topology.

11.5 Second-Side Resonance

Up until this point, in the topologies of quasi-resonant switching power supplies the resonant elements have been directly connected together in classic series tank circuit arrangements. They work quite well in this arrangement, but the largest obstacle that the designer must overcome is not the schematic design. As described in Section 11.6, the parasitic elements within the physical design of the components or layout can make an excellent schematic design into a mediocre power supply. Much of the designer's time is spent locating and defining these parasitic influences and minimizing their effects. The experienced designer will look for ways to recover the loss and turn it into a benefit for the supply. This is what can be done with two of the more annoying parasitic elements within transformer-isolated quasi-resonant converters: the leakage inductances and the interwinding capacitances of the transformer. These elements are bothersome because they form a high-frequency series tank circuit in series with all the windings of the transformer and cause high-voltage spikes to appear during voltage and current transitions. This causes increased RFI radiation and may damage some of the surrounding parts. Within PWM supplies, there was not a lot the designer could do other than place a snubber or a clamp across the primary winding. Within quasi-resonant supplies, though, there is a way to lump these parasitic elements inside other components, thus eliminating their independent effects on the operation of the supply.

As one can see in Figure 11.12, when the resonant elements are placed only on the primary side of the transformer, the leakage inductances and interwinding capacitances appear as independent parasitic elements in series between the resonant capacitor and the output $L–C$ filter. Here they can cause spikes that ring at their natural resonance

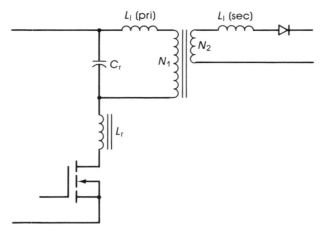

Figure 11.12
Leakage inductances within a quasi-resonant converter.

frequency, especially when the power switch and rectifier turn off. By employing a modeling technique for the transformer and "pushing" the resonant capacitor through the transformer to the sencondary, we see that the leakage inductances and interwinding capacitances now become elements within the series resonant tank circuit. Their values can now be added to the values of the resonant inductor and capacitor. The result is a welcomed absence of the voltage spikes on all the windings.

In order to "push" the resonant capacitor through the transformer, an adjustment must be made to the value of the capacitor by exactly the square of the turns ratio of the primary to the secondary windings. The expression for calculating the equivalent resonant capacitor on the secondary is

$$C_r(\text{sec}) = \left(\frac{N_{\text{sec}}}{N_{\text{pri}}}\right)^2 C_r(\text{pri}) \tag{11.5}$$

What this equation is showing is that there is a "phantom" resonant capacitor on the primary that is the impedance of the actual resonant capacitor when it is reflected back to the primary winding, thus maintaining the series resonant tank circuit topology on the primary. The values of the leakage inductances and interwinding capacitances can sometimes be comparable to the values of the resonant elements, so some readjustment in the values of the resonant elements may be necessary.

There is one small area of difficulty with the use of this technique in the multiple-output quasi-resonant switching regulator: the resonant capacitor should ideally be across all the secondaries. This is because any secondary that does not have a capacitive impedance placed across it will reflect a resistive impedance back to the primary and alter the resonant characteristic of the tank circuit. This may result in non-zero-current or voltage switching. Second-side resonance can be difficult to employ where it is required that all the outputs be DC isolated from one another. In this case, the designer may have to divide the value of the desired primary capacitor by the number of secondaries and then "push" each reduced value through the transformer to its respec-

Figure 11.13

Second-side resonance with isolated secondaries.

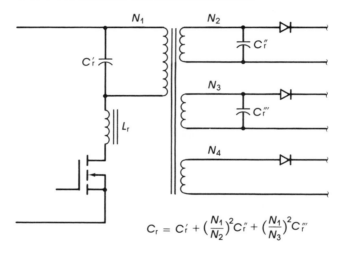

$$C_r = C'_r + \left(\frac{N_1}{N_2}\right)^2 C''_r + \left(\frac{N_1}{N_3}\right)^2 C'''_r$$

Figure 11.14

Second-side resonance with multiple stacked secondaries.

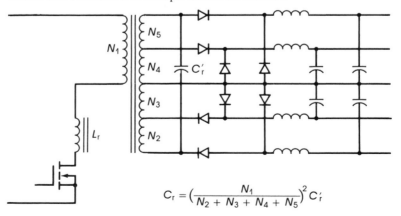

$$C_r = \left(\frac{N_1}{N_2 + N_3 + N_4 + N_5}\right)^2 C'_r$$

tive secondary. This would mean that each winding would have a resonant capacitor placed across it. Another method might be to split the resonant capacitor between the primary and secondary and push only half its value through the transformer to only the major secondaries as shown in Figure 11.13. This would tend to minimize the effects of the nonresonant secondary and still include the major leakage inductances within the tank circuit. Ideally, though, a single multiple-tapped secondary winding with the resonant capacitor across the entire winding (end-to-end) is best accomplished as shown in Figure 11.14.

The second-side resonance technique can result in a noticeable improvement in the performance of a quasi-resonant converter.

11.6 Effects of Parasitic Elements within High-Frequency Supplies

Parasitic elements are those electrical characteristics in the supply that are not intended as the primary function of the components used within the circuit but are nonetheless contributed by the physical construction and layout of the components. Many of these parasitic elements within the supply could be ignored in the PWM family of switching power supplies when their frequency of operation is less than 50 kHz because their overall effect on the power supply was minimal. At higher frequencies of operation, these effects become much more significant in the role they play in the efficiency of the power supply. Some are resistive in nature, such as the ESR within the capacitors and the eddy-current and hysteresis losses within the magnetic components. These dissipate real (nonreactive) power, which contributes to the heat generated within the supply. Some are capacitive in nature, that is, do not generate heat but form an AC current path that bypasses the main power transformation current path such as MOSFET gate capacitance and output capacitance. Finally, there are those parasitic effects that are magnetically induced that can couple into resistive losses such as skin effects in wires, and those that degrade the performance of components within the circuit, such as leakage inductance and interwinding capacitance. The effects of all these parasitic elements are frequency-dependent and radically increase as the frequency of operation increases. The losses contributed by parasitic effects can, if not properly attended to, contribute as much as 30 percent of the supply's total losses. Many can be reduced by the proper selection of components; others can be harnessed and used inside the power path, thus transforming a loss into usable energy. Within the realm of resonant switching power supplies, the designer must possess a practical appreciation of the parasitic effects to produce an efficient, well-designed switching power supply.

11.6.1 Transformer- and Inductor-Centered Parasitic Effects

The magnetic components such as transformers and inductors exhibit or cause two forms of parasitically caused losses. These are core-centered losses and air-coupled losses. The core-centered losses are those that you may have been familiar with in PWM power supplies. The air-coupled losses are often ignored in low-frequency PWM supply design. The transformer can contribute or cause up to half of the parasitic losses within the supply. So a good understanding of the transformer's influence on these losses should be pursued.

Core-Centered Parasitic Losses

These losses are contributed by the eddy-current losses and hysteresis losses of the core. These losses are highly dependent on the core material used within the transformer or inductor. Unfortunately, at present, there is not one single core material that satisfies all the demands placed on it for high-frequency power conversion purposes. First, it is desired that the B–H characteristic be as narrow as possible to reduce the hysteresis losses. Manganese–zinc materials possess such a narrow B–H characteristic. But these materials have a low volume resistivity, which promotes eddy-current losses at high frequency. Second, volume resistivity must be reasonably high to discourage eddy currents at high frequencies. Nickel–zinc ferrites provide this, but they have wider B–H characteristics, which promotes hysteresis and residual losses. The following expression describes the losses within the core:

$$\frac{R_{ac}}{\mu L} = a \cdot B_{max} \cdot f + cf + ef^2 \qquad (11.6)$$

where R_{ac} is the core loss resistance (in ohms), μ the permeability, L the inductance (in henries), a the hysteresis loss coefficient (published), c the residual loss coefficient (pubished)), e the eddy-current loss coefficient (published), B_{max} the maximum operating flux level (from application), and f the frequency (in hertz). The first term within the core-loss-summed equation is the hysteresis loss, the second is the residual loss, and the third is the eddy-current loss. Their respective loss coefficients are completely determined by the core material chosen for the application. So core material selection should play a prominent role in the transformer design. Note that the hysteresis losses are proportional to the maximum excursion of the flux during normal operation. This forces the designer to design the transformer with a lower B_{max}. Within PWM supplies, the designer typically sets B_{max} at one-half of the saturation flux density (B_{sat}). Within resonant supplies, though, B_{max} is typically set no higher than 10 to 15 percent of the saturation flux density. This presents some difficulties to the designer because more turns would have to be added to the transformer to reduce B_{max} and the core size would have to be increased because of both the increased windings and the less effective utilization of the core material. The last term in the core loss equation is the eddy-current loss within the core. As one can see, its value is proportional to the square of the frequency, so its significance increases greatly with the operating frequency of the supply. The resistivity of the core material affects the quantity of circulating eddy currents within the core by adding series resistance to its circulating current path. Hence, a material that possesses a higher volume resistivity helps in discouraging eddy currents. Some of the ferrite core materials on the market today that are targeted at high-frequency power conversion are 3F3 from Ferroxcube, H7F from TDK, and K or R material from Magnetics. In summary, the core selection and flux density limits chosen play a major role in controlling the internal core losses.

Winding-Centered Losses

A second major loss within the transformer is the skin effect within the windings themselves. The skin effect is caused by the existence of high currents at high frequencies. A large magnetic field is produced within the wire that is normal to the surface of the wire. This, in effect, "pushes" the current from the center of the wire to the wire's surface. This reduces the effective cross-sectional area of the wire that is available for current-carrying purposes and hence increases the resistance of the wire. The result is a greater winding loss than predicted by using a DC model for the winding, and it is more pronounced within the higher current windings. The expression for the skin effect is

$$\delta = \frac{0.066}{\sqrt{f}} \quad \text{(meters)} \qquad (11.7)$$

(See also Fig. 11.15.)

Several solutions can be used to reduce this effect. First is the use of Litz wire. Litz wire is a woven bundle of small-diameter insulated wires. This provides much more surface area, which promotes current

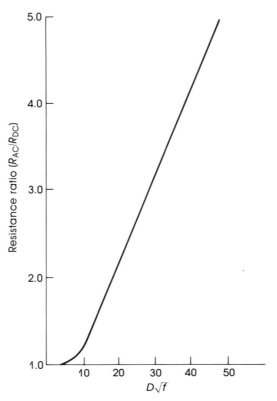

Figure 11.15

Resistance ratio (R_{AC}/R_{DC}) for solid copper wire.

sharing between the wires. The sum of the individual cross-sectional areas of the wires equals the cross-sectional area of the necessary equivalent solid conductor wire. The benefit of Litz wire can disappear if too many wires are contained within the bundle. The optimal number is about 5. This is partly because the high current windings usually have only a few turns, making it unlikely that all the wires within the larger Litz wire bundle emerge to the surface of the bundle. Another method is to use a flat foil conductor for the low-voltage, high-current windings within the transformer. The thickness of the foil is no more than two skin depths thick at the fundamental frequency of the current waveform. This better ensures that all the wire's cross-sectional area will be used for current flow. Using one or both of these techniques in the design can significantly reduce the winding loss due to the skin effect.

Another phenomenon that occurs within the windings is the existence of parasitically induced eddy currents caused by areas of high magnetic field strength. These areas of high magnetic field strength are typically caused by any gap contained in the core. Some of the lines of flux at the gap are bowed out away from the center of the gap in what is called the *fringe effect*. These lines of flux pass through the windings and induce eddy currents to flow within them. This, of course, causes a resistive-type loss within the windings, which results in still more heating of the winding. This is where the trade-off in wire type and winding placement comes into play. Eddy currents are more easily induced within conductors that have larger cross-sectional dimensions. The foil windings have a large surface area dimension, which gives them the greatest propensity for the induction of eddy-current flow. Solid magnet wire has a fairly large diameter dimension that also can permit the induction of eddy currents within them. Litz wire offers the best resistance to the induction of eddy currents because of the small diameter of each of these wires. This is where three trade-offs should be considered: (1) dielectric isolation between windings, (2) the degree of magnetic coupling needed by each winding to the core and to the other windings to minimize leakage inductance, and (3) which windings are better implemented using Litz or foil wires. Ideally, a winding composed of Litz wire should be placed adjacent to the core, followed by wires of ascending cross-sectional dimensions. This goes contrary to the normal practice of interleaving windings where half the primary is wound first, followed by the secondaries, and then the remainder of the primary last. Dielectric isolation is achieved by placing two layers of Mylar tape between the primary and secondary windings. Unfortunately, the windings most needing the Litz wire are the low-voltage, high-current secondaries. An additional layer of Mylar tape is required to ensure the required dielectric isolation between the primary and secondaries (see Fig. 11.16). It also decreases the magnetic coupling of the primary to the core, thus increasing its leakage inductance. This can result in the introduction of spikes into the primary's switching waveforms, thus requiring the addition of a clamp. Alternatively, Litz wire could be used for the primary, but because of its larger diameter and the fact that the primary usually requires the largest number of turns, the core size may have to be increased as a result. Obviously, not all the conditions can be satisfied, so some experimenting with the winding arrangement may be necessary to minimize the winding losses.

11.6.2 Layout- and Component-Dependent Parasitic Losses

These parasitic losses are distributed throughout the supply. Many of them are capacitive losses or magnetically induced resistive losses.

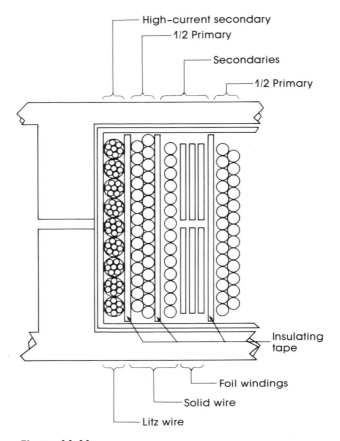

High–current secondary
1/2 Primary
Secondaries
1/2 Primary
Insulating tape
Foil windings
Solid wire
Litz wire

Figure 11.16

Method of minimizing winding eddy currents and maintaining good coupling dielectric isolation.

These parasitic losses are difficult to identify and quantify, but sometimes the designer can minimize their effects.

Capacitive Parasitic Effects

These parasitic equivalant elements tend to be centered about the power switch, which is typically a power MOSFET. The power MOSFET has three major capacitive parasitic elements within its high-frequency model: C_{gs} (gate-to-source capacitance), C_{dg} (drain-to-gate capacitance), and C_{ds} (drain-to-source capacitance). (see Fig. 11.17). Although these specific elements are not specified within a typical data sheet, they can be derived from the information given.

The first major capacitive loss within the MOSFET is caused by the gate-to-source capacitance. For each transition from on to off and vice

Figure 11.17

Gate drive loss (P_B = power lost in creating the gate drive voltage).

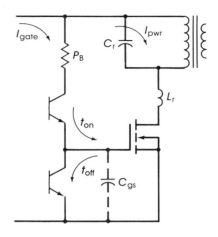

versa, this gate capacitor must be charged and then discharged. Although there is no power loss within the MOSFET itself, it does create heat-producing power losses in the gate drive circuit. The amount of energy that is driven into and sunk out of the gate varies with the PRF of the control signal and is given by

$$Q = \tfrac{1}{2}C_{gs} \cdot V_g^2 \cdot f \qquad (11.8)$$

One may be alarmed at the amount of power that is required to drive the gate at high frequencies, but one also must consider that it is in proportion to the output power. This means, for example, that it may cost the supply 3 percent in degraded efficiency at light loads but also cost the supply 3 percent at high loads. To minimize this penalty in lowered efficiency, the designer should consider how best to create the 10- to 12-V gate drive voltage. This consideration is completely analogous to the considerations given to the design of fixed base drive circuits discussed earlier for PWM supplies, and the amount of power required by the gate at these frequencies can be comparable. So if the elementary approach of dropping the input voltage to 10 V via a linear regulator is taken, the designer will be penalized severely in overall supply efficiency. A more moderate approach is the bootstrap starting circuit where the supply is started by a linear regulator from the input voltage until the outputs come up to the rated values. Then a + 12-V output provides all the gate drive current needed by the MOSFET. This improves the efficiency in generating the + 10-V gate drive voltage from the range of 10 to 30 percent to 80 percent. This can improve the overall efficiency of the supply by 5 percent. Ideally, it would be desirable to derive the gate drive current from the current flowing through the primary similar to proportional base drive circuits in PWM supplies, but no clean approach has yet been developed. This would turn a penalty into a benefit by including the gate drive current in the main power path and contributing to the output power instead of being shunted around it. Needless to say, the designer should not ignore the design of this area of the circuit.

Another capacitive parasitic loss is associated with the drain-to-source capacitor of the MOSFET and an often ignored parasitic capacitor that is formed between the case of the MOSFET (or transistor) and the heatsink. The later capacitor is a fixed value but can vary from supply to supply depending on the insulator thickness and the material plus the torque applied to the mounting screws (see Fig. 11.17). The C_{ds} of the MOSFET, which parallels the case-to-heatsink capacitor in the model (Fig. 11.18), can be very nonlinear depending on how the MOSFET is being operated. When the MOSFET is on, the C_{ds} is a low-value fixed capacitor in the range of tens of picofarads. When the MOSFET is in the off condition the value of C_{ds} increases to many hundreds of picofarads (see Fig. 11.19). Both of these capacitances can present a problem in some configurations of zero-current, quasi-resonant switching regulators in that there are rapid transitions in voltage across the drain and source terminals of the MOSFET and this stored charge is dissipated. The "off" value of C_{ds} does not present a significant problem to the supply since during the turn-on transition the value of C_{ds} drops from hundreds to tens of picofarads and the stored charge from the plummeting-value capacitor (Fig. 11.20) is summed into the on-state conduction current entering the resonant elements. The effects of the remaining fixed parasitic capacitors can be utilized inside a zero-voltage, quasi-resonant supply by simply including their values within the resonance capacitor. Another method to reduce their effects

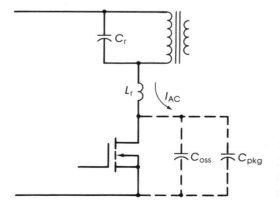

Figure 11.18
Parasitic capacitive losses associated with the drain and source of the power MOSFET.

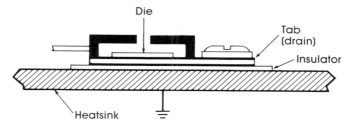

Figure 11.19

Parasitic capacitor between a power case and the heatsink.

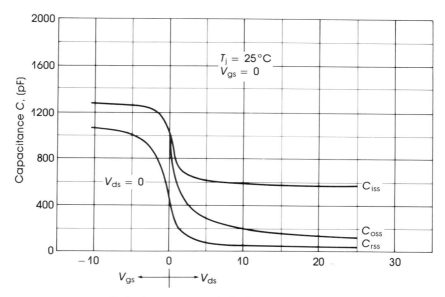

Gate-to-source or drain-to-source voltage (volts)

Figure 11.20

Capacitance variation for MTP8N20-Ref (courtesy of Motorola, Inc.).

is to use a heatsink insulator that has an embedded conductor sheet within it. This will effectively halve the case-to-heatsink capacitance by placing two capacitances of identical value in series. These are available from several manufacturers.

Magnetically Induced Losses

These losses are caused by localized, high-field-strength magnetic fields adjacent to the transformers and inductors within the supply. These losses are the most difficult to locate because they are so hidden within the physical layout of the power supply. The stray magnetic fields that escape from the transformer and inductors cause eddy currents to flow in any metal surface close to the magnetic elements. These surfaces can be ground planes or metal support members or can even exist within capacitors. To minimize these effects, locate the magnetic elements as far away from the metal support members and large capacitors as physically possible, and use a cross-hatched ground plane around the magnetic elements if a ground plane is needed in the area.

12

Switching Power Supply Design Examples

The following design examples were selected to give the reader a good insight into the design considerations involved in various switching power supply topologies. Hopefully, these will provide a basic "template" for any design that is attempted by the reader and will serve as a useful resource toward this end.

12.1 A Low-Cost, Low-Power Flyback Converter

12.1.1 Design Specification

Input voltage range:	12 to 28 V DC
Outputs (rated):	+ 12 V DC at 0.5 A
	− 12 V DC at 0.5 A

The switching power supply is to power some drivers that have intermittent load demands. The loads can vary from 0.1 to 0.5 A.

The supply is to be built on a PCB as part of a system design and has a 4-in.^{2} space allocation. The board will be connected to the power source by means of two connectors, each having 10 mΩ of contact resistance and 1.5 ft of #18-AWG wire. It also must cost no more than $15 in parts cost.

12.1.2 Predesign Considerations

1. The total source resistance due to the method of wiring within the system is

$$2(0.10 \ \Omega) + 1.5 \ \text{ft}[(6.381 \ \Omega/1000 \ \text{ft})/1000] = 29.6 \ \text{m}\Omega$$

This, in addition to the series inductance, makes it necessary that the input bulk filter capacitor be a good, low-ESR type.

2. The full load output power is

$$P_{out} = (12 \text{ V} \times 0.5 \text{ A}) + (12 \text{ V} \times 0.5 \text{ A}) = 12 \text{ W}$$

3. The estimated input power is

$$P_{in} = P_{out}/(\text{eff}) = 12 \text{ W}/0.75 = 16 \text{ W}$$

4. The estimated average input current is

$$I_{av} \text{ (low-line)} = P_{in}/V_{in}(\text{min}) = 16 \text{ W}/12 \text{ V} = 1.33 \text{ A}$$
$$I_{av} \text{ (high-line)} = P_{in}/V_{in}(\text{max}) = 16 \text{ W}/28 \text{ V} = 0.57 \text{ A}$$

5. The estimated maximum peak current is

$$I_{pk} = \frac{2P_{out}}{V_{in}(\text{min}) \times \delta_{max}} = 2(12 \text{ W})/(12 \text{ V} \times 0.5)$$

$$= 4.8 \text{ A}$$

6. From the input average current value, one can select the wire size of the primary to be #20 AWG for an average current density of 500 circular mils per ampere.

12.1.3 Transformer Design

1. The maximum inductance needed for the primary winding is

$$L_{pri} = \frac{V_{in}(\text{min}) \cdot \delta_{max}}{I_{pk} \cdot f} \qquad (12.1)$$

$$= \frac{12 \text{ V} \cdot 0.5}{4.8 \text{ A} \cdot 40 \text{ kHz}}$$

$$= 25 \text{ }\mu\text{H}$$

2. Select a core type and material. Molybdenum–permalloy (moly-permalloy) toroid cores offer the best winding-to-winding and core-to-winding coupling of various core types. It is also self-gapped. Unfortunately, toroids are more expensive to wind when compared to the bobbin-style cores in manufacturing labor costs. But with the better coupling characteristics, perhaps the toroid will preclude the need for adding a snubber to the circuit later. The molypermalloy toroid is selected.

3. Determine the toroid size. For molypermalloy cores, Magnetics, Inc. uses a method that determines the power the core must store to de-

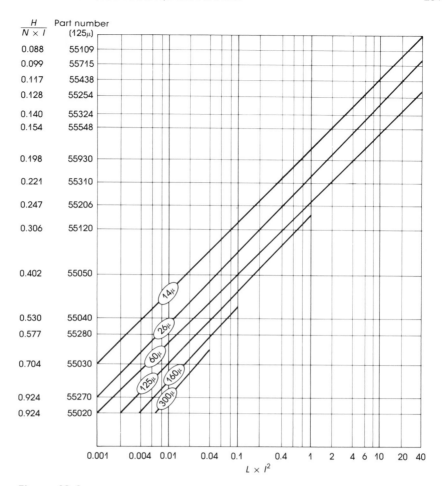

$\dfrac{H}{N \times I}$	Part number (125μ)
0.088	55109
0.099	55715
0.117	55438
0.128	55254
0.140	55324
0.154	55548
0.198	55930
0.221	55310
0.247	55206
0.306	55120
0.402	55050
0.530	55040
0.577	55280
0.704	55030
0.924	55270
0.924	55020

Figure 12.1

DC bias core selector chart (L = inductance with DC bias in millihenries; I = DC current in amperes). (Courtesy of Magnetics, Inc.)

termine the size and permeability of the core. This is done as follows:

$$LI^2 = (0.025)(4.8 \text{ A})^2 \quad (L \text{ is in millihenries}) \qquad (12.2)$$
$$= 0.58$$

Next, refer to Figure 12.1 and locate 0.58 on the x axis (horizontal) and proceed vertically until the first plot is intersected. That curve gives the needed permeability and the core part number is read from the y axis (vertical). Use the next higher core part number if it falls between curves. In this case it indicates that the core part number is 55120, which is a 0.65-in.-diameter core with a permeability of 125.

4. Determine the number of turns needed for the primary winding. Referring to the data sheet, note that this core exhibits 72 mH per 1000 turns. You can then determine the turns needed to develop 25 μH by

$$N_{\text{pri}} = 1000 \sqrt{\frac{L_{\text{pri}}}{L_{1000}}} \tag{12.3}$$

$$= 1000 \sqrt{\frac{0.025}{72}}$$

$$= 18.6 \text{ turns, rounded off to } 19 \text{ turns}$$

5. The number of turns needed for the secondaries is

$$N_{\text{sec}} = \frac{(V_{\text{out}} + V_{\text{D}})(1 - \delta_{\text{max}})N_{\text{pri}}}{V_{\text{in}}(\text{min}) \cdot \delta_{\text{max}}} \tag{12.4}$$

$$= \frac{(12 \text{ V} + 1 \text{ V})(1 - 0.5)(19 \text{ turns})}{12 \text{ V} \cdot 0.5}$$

$$= 20.5 \text{ turns, rounded off to } 20 \text{ turns}$$

It is better to round the turns downward within flyback-mode supplies.

6. The wire size needed for the secondaries is #24 AWG, once again using a current density of 500 circular mils per ampere.

Comment on the Physical Design of the Transformer

Since all the voltages are less than 42.5 V, there is no need to meet any dielectric withstanding voltage tests required by the safety approval agencies. This allows the designer to multifilar-wind all the windings on the transformer in order to optimize the interwinding coupling. This is done by cutting three lengths of magnet wire and twisting them together before winding them onto the core. For wire of this size a twist "pitch" of three twists per inch is about right. One caution: If the twist is too tight, the insulation could crack and the dielectric isolation could be lost, resulting in short-circuited turns.

When separating the ends, please pay special attention to the polarity of the windings. The polarity of the windings should be rechecked at the time the transformer is placed in the circuit and power is first applied.

12.1.4 Selecting the Semiconductors

Selecting the Output Rectifiers

In flyback switching power supplies, it is a good idea to select the current of the output rectifiers to be greater than 1.5 times the average output current since the peak currents are much greater than the average currents:

$$I_D > 1.5(I_{av}) = 1.5(0.5 \text{ A}) = 0.75 \text{ A}$$

The voltage rating of the rectifier should be greater than the total worst-case voltage the diode will see in its operation. This is

$$V_D > V_{out} + \frac{N_s}{N_{pri}} V_{in}(\text{max})$$

$$V_D > 12 \text{ V} + (20/19)(28 \text{ V}) = 41.5 \text{ V}$$

For flyback switching power supplies the forward recovery time of the output rectifiers significantly influences the size of the spike when the power switch turns off. So it is advantageous to select the fastest diode possible. In this case we are still within the voltage ratings of Schottky barrier diodes. The next best choice would be ultra-fast-recovery diodes. The choices can then be

MBR050   Schottky 50 V at 1 A

or

MUR110   ultra-fast recovery, 100 V at 1 A

Selecting the Power Switch

It was decided to use a power MOSFET as the power switch, although a bipolar power transistor could just as easily be used. The transistor would exhibit a higher switching loss and the power needed to drive it would be higher. The voltage rating of the MOSFET is chosen by estimating the worst-case voltage it would see during its operation plus a safety margin. This is done as follows.

$$V_{ds} > V_{in} + \frac{N_{pri}}{N_{sec}} (V_{out} + V_D) + V_{spike}(\text{est})$$

$$V_{ds} > 28 \text{ V} + (19/20)(12 \text{ V} + 1) + V_{spike}$$

$$V_{ds} > 40.3 \text{ V} + V_{spike}$$

From experience, this type of multifilar wound toroid produces a spike of approximately 35 V. So $V_{ds} > 75.3$ V. For an additional margin let's use 100 V. The current rating of the MOSFET is dictated by the FBSOA curve of the MOSFETs. Typical MOSFETs can withstand 3 times their average current rating in a nonrepetitive situation. Flyback converters have peak current of 4 to 5 times the average input current. So a good conservative current rating is

$$I_D > 1.5 I_{in}(\text{av}) = 1.5(1.33 \text{ A}) \approx 2 \text{ A}$$

The selected MOSFET is an MTP2N10. If a lower conduction loss is desired, a higher current MOSFET should be used at a very little increase in parts cost.

12.1.5 Design of the Controller

This example will illustrate two low-cost solutions to the same application. The first will utilize an MC34063, fixed on-time, variable off-time voltage-mode controller. The second approach will utilize the UC3843 current-mode controller.

MC34063 Implementation

The current-sensing resistor is

$$R_{sc} = 0.33/5 \text{ A} = 0.066 \text{ }\Omega$$

The timing capacitor is

$$C_t = (0.00004)(T_{on}) = (0.00004)(10 \text{ }\mu\text{sec})$$
$$= 400 \text{ pF} \quad \text{(use 470 pF)}$$

The resistor divider, which may be an iterative process in order to arrive at common resistor values, is

$$V_{ref} = 1.25 \text{ V DC}$$

Select the sense current of 1 mA.

$$R_{lower} = 1.25 \text{ V}/0.001 \text{ A} = 1.25 \text{ k}\Omega$$
$$R_{upper} = (12.0 \text{ V} - 1.25 \text{ V})/0.001 \text{ A} = 10.75 \text{ k}\Omega$$

Better values turned out to be

$$R_{lower} = 1.8 \text{ k}\Omega$$
$$R_{upper} = 15 \text{ k}\Omega$$

The startup resistor from the input line to pin 8 of the IC is

$$R_s = 12 \text{ V}/0.005 \text{ A} = 2.4 \text{ k}\Omega \quad \text{(make 2.2 k}\Omega\text{)}$$

This is a risky startup design in that the actual drive voltage for the MOSFET is not there during the first couple of pulses of the supply. The startup current flows through the base–emitter diode of the output transistor in the controller to drive the gate of the MOSFET. After the first couple of pulses, the output drive increases in voltage and supplies the full drive. Since the controller turns off after approximately 100 μsec, the power loss within the MOSFET is easily handled.

The output filter capacitors are

$$C_o = \frac{I_{out}(max) \cdot T_{off}}{V_{ripple}} = \frac{I_{out}(max)}{V_{ripple} \cdot f} \tag{12.5}$$

$$C_o = 125 \ \mu F$$

Since this is a variable-frequency mode of control, it is a good idea to increase the size of output capacitors.

Make $C_o = 220 \ \mu F$ at 20 V tantalum

Finally, to make the MOSFET turn off quickly, the active pull-down circuit is added to the gate of the MOSFET. This will force the MOSFET to turn off in less than 100 nsec. The resulting schematic is shown in Figure 12.2. If good grounding practices are followed and the transformer is wound as described above, then its performance is quite satisfactory. The measured efficiency is 81 percent at the rated loads and 89 percent at minimum loads.

UC3843 Implementation

Calculate R_t and C_t; from the graphs on the data sheet (40 kHz at 25 percent deadtime):

$$R_t = 4.7 \ k\Omega$$

$$C_t = 0.01 \ \mu F$$

The value of the startup resistor should be

$$R_{startup} = 12 \ V/0.005 = 2.4 \ k\Omega \quad (make \ 2.2 \ k\Omega)$$

The MOSFET startup energy is stored in the 10-μF capacitor and will provide enough energy when the undervoltage lockout circuit allows starting of the circuit.

From the MOFSET databook, the current-sensing resistor for the current-sensing MOSFET average current trip threshold 0.7 V (V_{sense}) is

$$R_{sense} = \frac{- V_{sense} \cdot R_{DM}}{V_{sense} - I_{pk} \cdot R_{DS(on)}} \tag{12.6}$$

$$= \frac{-0.7 \ V \cdot 288 \ \Omega}{0.7 \ V - (4.8 \ A)(0.25 \ \Omega)}$$

$$= 387 \ \Omega$$

For $R_{sense} = 387 \ \Omega$ the closest lower value is 360 Ω. (R_{DM} and $R_{DS(on)}$ are from the data sheet.)

A light filter is added to the current-sensing resistor to reduce the

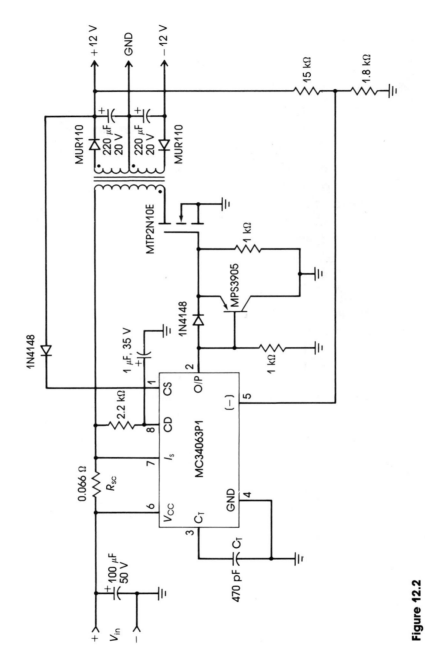

Figure 12.2

A low-cost, 12-W flyback converter (V_{in}: 12 to 28 V DC).

leading-edge spike that sometimes appears on the current waveforms. This spike causes erratic operation of the current-mode controller.

The voltage divider is calculated as follows: V_{ref} into the error amplifier is 2.5 V. Select the branch current at 0.926 mA:

$$R_{lower} = 2.5 \text{ V}/0.926 \text{ mA} = 2.7 \text{ k}\Omega$$
$$R_{upper} = (12 \text{ V} - 2.5 \text{ V})/0.926 \text{ mA} = 10.25 \text{ k}\Omega$$

The error amplifier is compensated as follows (refer to section 10.5):

1. The value of A_{DC} is found from

$$A_{DC} = \frac{(V_{in} - V_{out})^2}{V_{in} \cdot \Delta V_c} \left(\frac{N_{sec}}{N_{pri}}\right) \qquad (12.7)$$

$$A_{DC}(\text{high}) = 15.5 \text{ dBV}$$

$$A_{DC}(\text{low}) = 0 \text{ dBV}$$

2. Maximum crossover frequency = $F_s/5$ = 8.0 kHz at the high-input line.
3. Filter pole (heavy load) = 302 Hz, filter pole (light load) = 6.03 Hz.
4. Gain of the compensated error amplifier at 8 kHz should be solved at the high input line condition since the control-to-output function exhibits its highest gain (G) at that point and will exhibit the widest loop bandwidth:

$$G(8 \text{ kHz}) = -20 \log(8000 \text{ Hz}/300 \text{ Hz}) + 15.5 \text{ dB} = 13 \text{ dB},$$
$$A = 4.5$$

5. Component values are

$$C_1 = 1/(6.28)(4.5)(8 \text{ kHz})(10.2 \text{ k}\Omega) = 430 \text{ pF}$$
$$R_2 = 4.5(10.2 \text{ k}\Omega) = 45.7 \text{ k}\Omega \text{ or } 47 \text{ k}\Omega$$
$$C_2 = 1/(6.28)(6.03)(47 \text{ k}\Omega) = 0.56 \text{ }\mu\text{F}$$

This is in anticipation of the worst-case zero frequency for the tantalum output capacitors being at 8 kHz.

The schematic for this implementation is shown in Figure 12.3. (*Note:* The point of joining the grounds is at the base of the current-sensing resistor, so all control grounds should be joined to the high current grounds at that point.)

12.1.6 Postpaper Design Note

The supply was found to be unable to supply the necessary power to the load. The current-sensing resistor was lowered to 330 Ω, and satisfactory operation was attained.

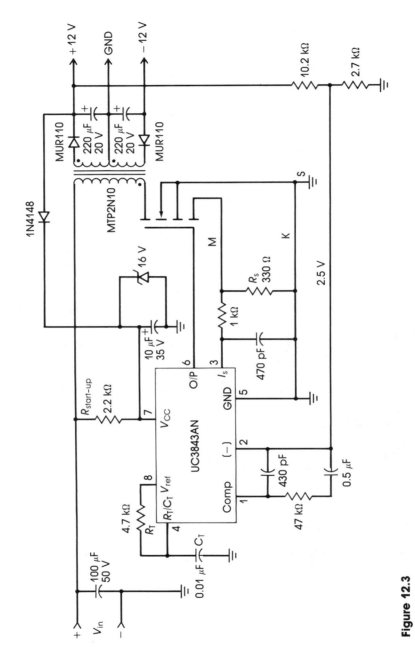

Figure 12.3

A low-cost, 12-W flyback converter using current-mode control (V_{in}: 12 to 28 V DC).

12.2 A 100-kHz, 50-W, Off-Line, Half-Bridge Switching Power Supply

12.2.1 Design Specification

Input voltage range: 90 V to 130 V and 205 V to 240 V RMS

Outputs (rated):

+5 V DC   at   5 A
+ 12 V DC   at   1 A
− 12 V DC   at   1 A

Output ripple voltage:

+5 V: 50 mV$_{p-p}$
+/− 12 V: 100 mV$_{p-p}$

The needed dielectric isolation from the primary to the outputs is 3750 V RMS to meet VDE specifications. A schematic of this power supply is shown in Figures 12.4 and 12.5 on pages 218 and 219.

12.2.2 Predesign Considerations

1. Output power:

$$P_{out} = (5 \text{ V x } 5 \text{ A}) + (12 \text{ V x } 1 \text{ A}) + (12 \text{ V x } 1 \text{ A}) = 49 \text{ W}$$

2. Estimated input power:

$$P_{in} = P_{out}/(\text{eff}) = 49 \text{ W}/0.75 = 65.3 \text{ W}$$

3. Rectified input voltage range: (worst case using voltage doubler for 115-V-AC line):

$$V_{in}(\text{low}) = 1.414(2)(90 \text{ V RMS}) - 25 \text{ V}_{\text{ripple}} = 230 \text{ V DC}$$
$$V_{in}(\text{high}) = 1.414(2)(130 \text{ V RMS}) - 25 \text{ V}_{\text{ripple}} = 343 \text{ V DC}$$

4. Estimated average input currents:

$$I_{av}(\text{low-line}) = P_{in}/V_{in}(\text{min}) = 65.3 \text{ W}/230 \text{ V} = 0.284 \text{ A}$$
$$I_{av}(\text{high-line}) = P_{in}/V_{in}(\text{max}) = 65.3 \text{ W}/343 \text{ V} = 0.190 \text{ A}$$

5. Estimated maximum peak current:

$$I_{pk} = 2.8P_{out}/V_{in}(\text{min}) = 2.8(49 \text{ W})/230 \text{ V} = 0.6 \text{ A}$$

12.2.3 Transformer Design

1. Select the core material and type. Reviewing the material specifications from the core manufacturers (Magnetics, Inc. in this case)

shows that a good material for core loss at 100 kHz and saturation flux density is the "F" material. The desired core configuration is an EC core. For bipolar flux, forward-mode applications, no gap is required.

2. Determine the optimum core size (using the equation from Magnetics, Inc. design information). At 20 to 40 kHz, the designer would normally operate the core at one-half of B_{sat} (2350 G). At 100 kHz it is advised to operate the core at a lesser value to keep the core losses to less than 2 percent of the total supply's losses. Lets use 1200 G:

$$W_a A_c = \frac{P_{out} \cdot I_D \cdot 10^8}{4(0.24)B_{max} \cdot f} \tag{12.8}$$

$$= \frac{(49\ \text{W})(500\ \text{cm/A})10^8}{4(0.24)(1200\ \text{G})(1 \times 10^5\ \text{Hz})}$$

$$= 21{,}267\ \text{circular mils}$$

Referring to the EC core data sheets, this $W_a A_c$ corresponds to an F42510-EC-00. Keeping in mind that this transformer must meet VDE dielectric requirements and that additional layers of insulating tape must be included within the windings, let's use the next size larger. This would make the part number F43515-EC-00.

3. Determine the number of turns required for the primary winding:

$$N_{pri} = \frac{V_{pri}(\text{nom}) \cdot 10^8}{4 \cdot f \cdot B_{max} \cdot A_c} \tag{12.9}$$

$$= \frac{(281\ \text{V}/2)10^8}{4(10^5\ \text{Hz})(1200)(0.904)}$$

$$= 32.3\ \text{turns, rounded off to 32 turns}$$

4. Determine the number of turns needed for the secondary windings. First determine the actual secondary voltages by adding the diode forward voltage drop to the output voltage and dividing by the maximum allowed duty cycle:

$$V_{sec}(\text{min}) = \frac{V_{out} + V_D}{\delta_{max}} \quad (\delta = \text{duty cycle}) \tag{12.10}$$

$$= \frac{5.0\ \text{V} + 0.5\text{V}}{0.95} = 5.79\ V_{min}\ \text{or 6 V}$$

This should be treated as the absolute minimum voltage that should be presented to the output L–C filters. Experience dictates that a margin in voltage should be added to these values to guarantee that the supply will remain within regulation at low line. So a good place to start is 6.0 V for the low-voltage output. To calculate the number of turns needed for the +5-V winding, use the transformer turns ratio equation:

$$N(5) = \frac{N_{pri} \cdot V_{sec}(\min)}{V_{pri}(\min)} \tag{12.11}$$

$$= \frac{(32 \text{ turns})(6 \text{ V})}{(230/2)\text{V}} = 1.67 \text{ turns} \rightarrow 2 \text{ turns}$$

Rounding this upward to the next full-turn (dictated by EC core construction), this becomes 2.0 turns. The actual peak output voltage with two turns is

$$V_{sec} = (230/2)(2/32) = 7.18 \text{ V}$$

This yields a volts-per-turn constant of 3.59 V per turn, corresponding to a maximum duty cycle of

$$\delta_{max} = 5.0 \text{ V}/(7.18 \text{ V} - 0.5 \text{ V}) = 74.9 \text{ or } 75 \text{ percent}$$

5. The turns needed for the +/− 12-V secondary windings are

$$V_{sec}(12) = (12 \text{ V} + 0.9 \text{ V})/0.75 = 16.5 \text{ V}$$

The turns then become

$$N(12) = 16.5 \text{ V}/(3.59 \text{ V/turn}) = 4.6 \text{ turns}$$

or 5 turns (EC cores must have an integer number of turns). The error is

$$V(12)(\text{actual}) = 5 \text{ turns}(3.59 \text{ V/turn})(0.75) - 0.9 \text{ V}$$
$$= 12.6 \text{ V (acceptable)}$$

6. The equivalent wire gauges needed for each winding are

$$
\begin{array}{ll}
\text{Primary (0.6 A average):} & \#22 \text{ AWG} \\
+12 \text{ V (1.0 A average):} & \#20 \text{ AWG} \\
-12 \text{ V (1.0 A average):} & \#20 \text{ AWG} \\
+5 \text{ V (2.0 A + 5 A):} & \#13 \text{ AWG}
\end{array}
$$

The +5 winding will form the bottom of the +/− 12-V windings, so the +5-V winding must support the current of the other windings. The +5-V winding will consist of smaller wires that will add up to the same cross-sectional area such as seven strands of a #22 AWG. Litz wire could also be used by first calculating the skin depth caused by the skin effect and then doubling this number to get the diameter of each wire within the Litz bundle. Finally, there should be enough combined wire area to support the current (at 500 circular mils per ampere).

7. Checking the amount of window area the wires occupy within the bobbin of this size core, we obtain

$$W_a = 32 \text{ turns}(807 \text{ cm}) + 2(2 \text{ turns})(5852 \text{ cm})$$
$$+ 2(5 \text{ turns} - 2 \text{ turns})(2)(1246 \text{ cm})$$
$$= 64{,}184 \text{ circular mils } (1 \text{ circular mil} = 7.85 \times 10^{-7} \text{ in.}^2)$$
$$= 0.05 \text{ in.}^2$$

This is fine—available area is 0.292 in.^{2}

12.2.4 Transformer Construction

To meet VDE requirements, there must be several layers of insulating tape between the primary and secondary windings. Also, to meet the creepage requirements, the windings are to be spaced 2 mm from the ends of the bobbin. The leads to and from the windings need Teflon insulating sleeving around the leads. The transformer will be wound using the interleaved winding technique, where the secondaries will be "sandwiched" between halves of the primary winding (see Fig. 7.2). Also, another winding will be added, which will provide power to the control IC and power switches. It will be associated with the primary side of the transformer. This should contribute very little to the wire area within the transformer since the controller draws very little current.

12.2.5 Design of the Output Filter Chokes

1. The +5-V filter choke is

$$L_{\min} = \frac{V_{in}(\max) \cdot T_{off}(\max)}{1.4 \cdot I_{out}(\max)} = \frac{(8.3 \text{ V})(1 \times 10^{-5}\text{sec})}{1.4(1 \text{ A})} \quad (12.12)$$

$$L_{\min} = 60 \ \mu\text{H}$$

2. The $+/-$ 12-V mutually coupled choke is

$$L_{\min} = 550 \ \mu\text{H for both outputs}$$

Refer to Section 12.1 for core selection and turns calculations. The mutually coupled filter choke is to be bifilar-wound to ensure that the number of turns for each choke is identical.

3. For the design of the gate drive transformer, I am going to use an MPP powder core, $\mu = 550$. The transformer will have a 1:1 turns

ratio. These cores are small because of the power. Let's start with a 0.310-in. diameter core.

$$N_{pri} = \frac{V_{pri} \cdot 10^8}{4 B_{max} \cdot f \cdot A_c}$$ (12.13)

$$= \frac{12 \text{ V} \cdot 10^8}{4(4000)(10^5 \text{ Hz})(0.00953)}$$

$$= 78 \text{ turns}$$

Checking to see if the window fillage is less than 50 percent, $W_a =$ 2(78 turns)(64 circular mils)/(18,200) $= 54.8$ percent. Let's try it.

4. The output filter capacitors are determined by

$$C_o = \frac{I_{out}(max) \cdot T_{off}(max)}{V_{ripple}(max)}$$ (12.14)

$$C_{out}(+5) = 1000 \ \mu F$$

Use four 220-μF/20-V tantalum capacitors.

$$C_{out}(+/-12) = 10 \ \mu F$$

Use tantalum capacitors with values of 33-μF/35-V for +12 and 47-μF/35-V for −12-V. The −12 V is unsensed, which can result in more ripple voltage.

12.2.6 Selection of the Semiconductors

Selection of the Power Switches

Because this supply is being designed to operate at 100 kHz, it is necessary to use power MOSFETs in order to minimize the switching losses.

$$V_{dss} > 1.3 V_{in}(max) = 1.3(367 \text{ V}) > 478 \text{ V}$$
$$I_d > 1.5 I_{in}(av) = 1.5(0.6 \text{ A}) > 0.9 \text{ or } 1 \text{ A}$$

The closest TO-220 package is MTP1N50, but with an 8-Ω R_{ds}(on), this yields a 2.3-W maximum saturation power loss. Going to the MTP2N50 cuts this loss in half.

Selection of the Output Rectifiers

+5 V: $V_r > 2V_{out} > 12$ V, $I_{fwd} > 5$ A, P/N is MBR735 (Schottky)
+/−12 V: $V_r > 2V_{out} > 24$ V, $I_{fwd} > 1$ A, P/N is MUR405 or
MUR410 (ultra-fast-recovery)

Selection of a Controller IC

Several features are dictated by the application. First, a two-channel controller is needed. The output drivers should be a totem-pole configuration, and it is desired that we use voltage-mode control. The IC should be capable of 100-kHz operation. Also, the IC should have an integral overcurrent amplifier, reference, soft-start, and deadtime adjustment. The IC that fits all these requirements is SG3526.

12.2.7 Designing the Bootstrap Startup Circuit

A bootstrap startup circuit is a small linear regulator that is required only for the initial startup of the power supply and for when the supply enters a foldback condition. After the supply's outputs come up to their specified output voltages, the linear regulator is no longer required and is cut off by the series diode. This removes a lossy means of generating the control IC supply voltage and MOSFET driving voltage. From then on the IC and MOSFET draw their power from an auxiliary winding.

1. The base resistor should have a value as high as possible since this will represent a steady-state loss:

$$R_b = V_{in}(\text{low})/I_b = 230 \text{ V}/0.5 \text{ mA}$$
$$= 460 \text{ or } 470 \text{ k}\Omega \quad (\text{power loss} = 0.115 \text{ W}_{max})$$

2. The collector resistor is

$$R_c = \frac{V_{in}(\text{low})}{I_c} = (230 - 10)/0.003 \text{ A}$$
$$= 73 \text{ or } 75 \text{ k}\Omega \ (0.66 \text{ W } P_d)$$

Although this resistor is used only momentarily during normal operation, this regulator can be reactivated if an overcurrent condition is encountered. So rate this resistor for steady-state operation at 1 W.

3. The transistor can be a low-power transistor, but it must withstand the full input voltage as its V_{ce}. A TIP50 looks good, with a V_{ce} of 400 V and an I_c of 1 A. This transistor is oversized, but there are no small signal transistors with this rating.

4. The zener is to produce an emitter voltage of 10 V. So adding the V_{be} drop yields about a 12-V zener. A 500-mW zener is more than enough. A 1N5241A will do.

12.2.8 Controller IC Associated Components

1. *Oscillator timing* R *and* C. From the graph within the data sheet
   $C_t = 2000$ pF, $R_t = 5.6$ kΩ
2. *Deadtime resistor.* $R_d \approx 0.5$ Ω
3. *Error amplifier voltage dividers.* With an optoisolator type of feedback, the reference must be divided to a lower voltage before it can be used by the error amplifier. Dividing by $\frac{1}{2}$ is convenient. In determining the reference divider (to the noninverting input) we want to draw 0.5 mA:

$$R_{total} = 5.0 \text{ V}/0.5 \text{ mA} = 10 \text{ k}\Omega, \quad R_{top} \text{ \& } R_{bottom} = 4.7 \text{ k}\Omega$$

The inverting input needs a voltage divider that provides a voltage below 2.5 V when the optoisolator is in cutoff. The reference used in the secondary voltage sense network must have at least 1 mA flowing through it to operate correctly. The optoisolator selected has a nominal 100 percent current transfer ratio ($I_{in} = I_{out}$). This renders the optoisolator contributed current 1 mA minimum.

$$R_{bottom} = 2.5 \text{ V}/(1 + 0.5 \text{ mA}) = 1.6 \text{ or } 1.5 \text{ k}\Omega$$
$$R_{top} = (5 \text{ V} - 2.5 \text{ V})/0.5 \text{ mA} = 5 \text{ or } 4.7 \text{ k}\Omega$$

Determine the sense current needed to be passed through the light-emitting diode (LED) of the optoisolator: (select the 4N35). The minimum current transfer ratio is 100 percent. This means that the maximum required sense current passed through the output voltage sense divider is

$$I_s\text{max} = I_{control}/C_{tr}(\text{min}) = 1 \text{ mA}/100 \text{ percent}$$
$$= 1 \text{ mA (at rated output)}$$

It is planned that an adjustable, temperature-compensated reference device such as the TL431 be used as the secondary reference. It has a minimum adjustable voltage of 2.5 V. The voltage drop across the TL431 and the LED (4N35) is

$$V_{ref}(\text{min}) = 2.5 \text{ V} + 1.5 \text{ V} = 4.0 \text{ V}$$

The sense current will be split between the +5- and the +12-V outputs in an 80:20 split, respectively. That makes the +5-V sense resistor:

$$R_s(+5) = (5.0 \text{ V} - 4.0 \text{ V})/0.8(1 \text{ mA}) = 1250 \text{ }\Omega \text{ or } 1.2 \text{ k}\Omega$$

The real resulting split is

$$1.0 \text{ V}/1.2 \text{ k}\Omega = 0.83 \text{ mA}$$

so the $+12$ V is 1 mA $-$ 0.83 mA $=$ 0.17 mA. The $+12$-V sense resistor is

$$R_s(+12) = (12 \text{ V} - 4.0 \text{ V})/0.17 \text{ mA} = 47 \text{ k}\Omega$$

12.2.9 Error Amplifier Compensation

Two-pole–two-zero compensation is optimum for a forward-mode converter. Please refer to Section 10.5 for the design equations.

1. The dominant (lowest-frequency) output filter pole is caused by the $+5$-V-output L–C filter (refer to Section 10.5):

$$f_p(+5) = 719 \text{ Hz}$$
$$f_p(+12) = 1292 \text{ Hz}$$
$$f_p(-12) = 1082 \text{ Hz}$$

For tantalum capacitors, the zero caused by the ESR of the capacitor times the value of the capacitor itself can be expected to be about 10 kHz, but to play it safe, a frequency of 8 kHz will be used.

2. The gain exhibited by the power network at DC is

$$A_{DC} = 20 \log\left(\frac{V_{in}}{V_{ramp}} \cdot \frac{N_s}{N_p}\right); \qquad V_{ramp} = 2.4 \text{ V} \quad (12.15)$$

Low line: $A_{DC} = 17.1 \text{ dB}$
High line: $A_{DC} = 20.6 \text{ dB}$

3. For voltage-mode control, forward-mode regulators, a two-pole–two-zero type of compensation is recommended (refer to Section 10.5). This will yield the best transient response.

4. The overall gain crossover frequency is approximately

$$f_{xo} = \frac{f_s}{5} = \frac{100 \text{ kHz}}{5} = 20 \text{ kHz} \quad (\text{maximum})$$

5. The gain needed to bring the control-to-output curve up to 0 dB at the overall crossover frequency is 29 dB (see Fig. 12.6 on page 219) (or a gain of 28.1).

6. The location of the two low-frequency zeros (f_z) will be at one-half the pole frequency of the output filter (the poles will coincide):

$$f_z = \frac{f_{L-C}}{2} = \frac{719 \text{ Hz}}{2} \approx 360 \text{ Hz}$$

7. The location of the first high-frequency pole is at the worst-case ESR zero frequency, which is approximately 8 kHz.

8. The location of the second high-frequency pole should be

$$f_{p2} = 1.5 f_{xo} = 1.5(20 \text{ kHz}) = 30 \text{ kHz}$$

9. The gain needed at the location of the two compensating zeros is

$$A_1 = A_2 + 20 \log\left(\frac{f_{z2}}{f_{p1}}\right) \tag{12.16}$$

$$= 29.2 \text{ dB} + 20 \log\left(\frac{360 \text{ Hz}}{8{,}000 \text{ Hz}}\right)$$

$$= 2.3 \text{ dB} \quad \text{(or a gain of 1.3)}$$

10. Find C_1 (R_{in} or $R_1 = 1.2 \text{ k}\Omega$):

$$C_1 = \frac{1}{2\pi A_1 \cdot f_{xo} \cdot R_1} \tag{12.17}$$

$$= \frac{1}{2\pi(2)(2 \times 10^4 \text{ Hz})(1.2 \text{ k}\Omega)}$$

$$= 3300 \text{ pF (rounded)}$$

11. Find R_2:

$$R_2 = A_1(R_1) = 1.3(1.2 \text{ k}\Omega) = 1.5 \text{ k}\Omega$$

12. Find R_3:

$$R_3 = \frac{R_2}{A_2} = \frac{1.5 \text{ k}\Omega}{28.1} = 51 \ \Omega$$

13. Find C_2:

$$C_2 = \frac{1}{2\pi f_{z2} \cdot R_2} \tag{12.18}$$

$$= \frac{1}{2\pi(360 \text{ Hz})(1.5 \text{ k}\Omega)}$$

$$= 0.29 \ \mu\text{F} \rightarrow 0.27 \ \mu\text{F}$$

14. Find C_3:

$$C_3 = \frac{1}{2\pi f_z \cdot R_1} \tag{12.19}$$

$$= \frac{1}{2\pi(360 \text{ Hz})(1.2 \text{ k}\Omega)}$$

$$= 0.36 \ \mu\text{F}$$

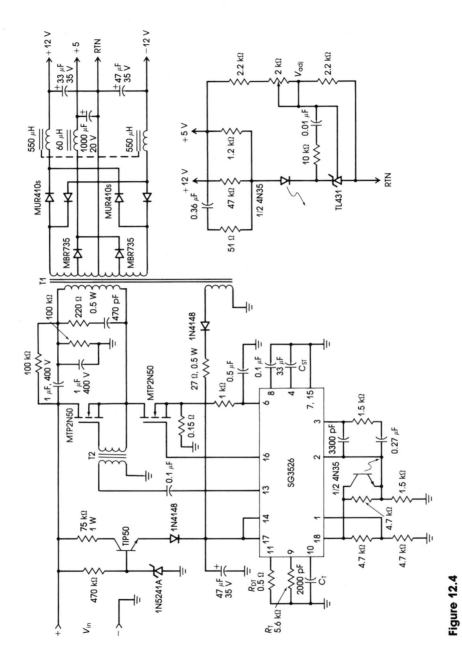

Figure 12.4

A 50-W, 100 kHz, PWM, half-bridge switching power supply.

218

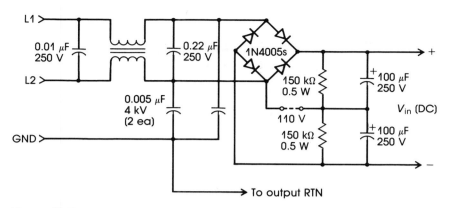

Figure 12.5
Input RFI/rectifier circuit for a 50-W half-bridge switching power supply.

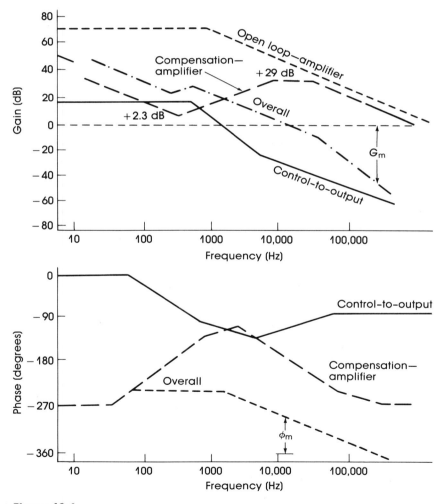

Figure 12.6
Compensation of the design example: Half-bridge voltage-mode control SMPS.

12.3 A 50-W, Parallel-Resonant, Half-Bridge, Quasi-Resonant Converter

This example demonstrates just how topologically similar quasi-resonant converters are to their PWM counterparts (see also Fig. 12.7 on page 226). Many of the same design procedures can be used as in the PWM half-bridge, except with some changes to the operating limits and the control method.

12.3.1 Additional Design Specifications

Resonant frequency: 1 MHz

Minimum control frequency (lightest load):   200 kHz

12.3.2 Predesign Considerations

1. The on-time of the power switch must be set in relationship to the resonant period of the tank circuit. Specifically, the on-time must end during the current ring-back period of the tank circuit when the intrinsic MOSFET diode is conducting. If the period of T_{on} is set to the center of the ring-back period, which is usually shorter than the positive resonant period, then T_{on} should be approximately 700 nsec. This will no doubt need to be adjusted after the prototype is built, as the resonant period of the tank circuit varies somewhat in proportion to the input voltage.

2. The minimum control period (or maximum control frequency) must be longer than the time required for the tank circuit to execute its half-sinusoid resonant period plus its time to empty the excess energy plus a deadtime for margin. This is to ensure that the tank circuit and transformer start from the zero-energy initial condition to meet zero-current switching requirements. Thus we obtain

$$T_{min} > 500 \text{ nsec} + {\sim}300 \text{ nsec} = 800 \text{ nsec}$$

At this point in the design, it is wise to initially add a wide margin to the deadtime of the minimum control frequency. This can always be changed after the prototype is working properly. lets use a 40 percent margin for the deadtime. Then the maximum control frequency is 900 kHz.

12.3.3 Transformer (Re)Design

1. Select the core material and type. There is ongoing research by core manufacturers in an effort to minimize core losses above 100 kHz. Two materials that work reasonably well at 1 MHz are the 3C85 (TDK) and the F (Magnetics) materials. The F material will be used for this example. Although a pot core would provide superior RF shielding of the windings, dissipating the heat is a problem. An EC core will be used instead. They usually have larger wire areas, which will make it easier to fit the insulating layers. A Faraday shield can be added for the shielding.

2. Try to arrive at a reasonable maximum operating flux density (B_{max}) for the transformer. Referring to the "core loss versus flux density" graph for the F material, in order to hold the core loss at the same level as the 100-kHz PWM converter in the design example of Section 12.2, the B_{max} will have to be set at 150 to 300 G. This has a "nonreducing" effect on the size of the core.

3. Select the size of the E core:

$$W_a A_c = \frac{P_{out} \cdot I_D \cdot 10^8}{B_{max} f_{max}} \tag{12.20}$$

$$= \frac{(49 \text{ W})(500)10^8}{(200 \text{ G})(9 \times 10^5)}$$

$$= 13{,}611 \text{ cm}^4$$

The core size that just exceeds this number is the F42510-EC-00. For the sake of being conservative, let's select the next size upward and worry about reducing the size later. So order F43515-EC-00.

4. The number of turns required for the primary can be found approximately by

$$N_{pri} = \frac{V_{pri}(nom) \cdot 10^8}{4.4 \cdot f \cdot B_{max} \cdot A_c} \tag{12.21}$$

$$= \frac{(325 \text{ V}/2)(10^8)}{4.4(9 \times 10^5 \text{ Hz})(200 \text{ G})(0.904)}$$

$$= 22.7 \text{ turns or 23 turns rounded off.}$$

(Assuming a sinusoidal excitation voltage applied to the primary winding.)

5. Determine the minimum number of turns needed for the secondaries. If we assume that the worst-case Q of the resonant tank circuit is 1,

we can determine the worst-case minimum peak voltage that will be needed on the secondary winding at the low input line voltage. This assumption is needed since the actual Q of the tank circuit is determined by the reflected impedance of the load multiplied by the turns ratio of the transformer, which we do not know yet. Also, if we assume that there is a pure one-half of a sinusoid waveform (which is a reasonable assumption), we can readily determine the peak voltage of the waveform.

(a)   The maximum possible duty cycle is

$$\delta_{max} = T_{on} \cdot f_{max} \qquad (12.22)$$
$$= (0.5 \ \mu sec) \cdot (900 \ kHz) = 0.45 \ percent$$

(b)   The required minimum peak voltage to keep the output within regulation is

$$V_{sec} = \frac{\sqrt{2}(V_{out} + V_D)}{\delta_{max}} \qquad (12.23)$$

$$= \frac{\sqrt{2}(5.0 + 0.5)}{0.45} = 17.3 \ V_{pK}$$

(c)   The number of turns needed for the $+5$-V secondary is

$$N_{sec} = \frac{(V_{out} + V_D)N_{pri}}{V_{pri} \cdot \delta_{max}} \qquad (12.24)$$

$$= \frac{(5 + 0.5)(30 \ turns)}{(230/2)(0.45)} = 2.4 \ turns$$

Rounding upward makes $N(+5 \ V) = 3$ turns. (Again assuming $Q = 1$ for the tank circuit; any higher Q would bring the regulation even more within the range of the control loop.)

(d)   The turns needed for the $+/-12$-V windings are

$$N(12 \ V) = \frac{(V_{out2} + V_{D2})N(1)}{(V_{out1} + V_{D1})} \qquad (12.25)$$

$$= \frac{(12 + 0.9 \ V)3 \ turns}{(5 + 0.5 \ V)}$$

$$= 7.4 \ turns$$

Rounding off makes $N(12 \ V) = 7$ turns.

6. The depth of the skin effect is given by

$$\delta_s = \frac{0.066}{\sqrt{f}} \quad (meters) \qquad (12.26)$$

The skin depth is then 2.6 mils. This means that ideally, to minimize wire losses, Litz wire in which the strand diameter averages 5 mils or less should be used. Foil windings should not exceed 5 mils either.

7. The construction of the transformer will be interleaved, similar to the previous example. The equivalent wire gauges are the same as in the previous example since the average currents have not changed. Skin effect now is a major concern, especially with the +5-V winding. The +5-V winding should be either 5-mil foil or Litz wire with 5-mil strands. All other windings can use Litz wire or consist of multiple strands of #24–#26 AWG wires to help reduce skin effect loss. Since the supply must meet VDE safety requirements, the primary windings will be insulated with tape from the secondary windings.

12.3.4 Designing the Resonant Tank Circuit

1. Many combinations of inductance and capacitance values have a resonant frequency of 1 MHz. It is important to balance the energy between the resonant L and C:

$$\tfrac{1}{2}L_r i^2 = \tfrac{1}{2}C_r V_r^2 = W \tag{12.27}$$

2. So rearranging Equation (12.27) for C_r and substituting $W = P_{out}/f$ (energy per cycle equals the average power divided by the frequency of operation) we obtain

$$C_r = \frac{2P_{out}}{V_{pri}^2 \cdot f} \tag{12.28}$$

$$= \frac{2(49\ W)}{(230/2)^2 \cdot (9 \times 10^5\ Hz)}$$

$$= 8233\ \text{pF or } 8200\ \text{pF}\quad \text{(closest standard value)}$$

3. Substituting this above value of C_r into

$$L_r = \frac{1}{C_r(2\pi f_r)^2} \tag{12.29}$$

$$= \frac{1}{(8.2 \times 10^{-9})(2\pi \cdot 1 \times 10^6)^2}$$

$$= 3.09\ \mu\text{H}$$

4. To reduce the need for a heavy snubber across the primary winding due to the leakage inductances, let's push the resonant capacitor through the transformer to the secondary. This will effectively sum

the primary and secondary leakage inductances into the resosnant inductor value. There may need to be some later adjustment to the resonant inductor value since it is difficult for the average engineer to measure the value of the leakage inductances. Neglecting this effect of the leakage inductances during the paper design, one determines the value of the second-side resonant capacitor by

$$C_r(\text{sec}) = C_r(\text{pri})\left(\frac{N_{sec}}{N_{pri}}\right)^2 \qquad (12.30)$$

$$= 8200 \text{ pF}\left(\frac{2(7 \text{ turns})}{23 \text{ turns}}\right)^2$$

$$= 3038 \text{ pF or } 2700 \text{ pF}$$

(rounding downward since the leakage inductances will increase the resonant capacitor)

This capacitor is to be placed across the entire secondary. That is between the $+/-$ 12-V secondary leads.

12.3.5 The Output Filter Capacitors

The output filter capacitors can be determined by the same method as in the PWM version, but that would yield values that are larger than required. Keep in mind that at the lowest operating frequency the lowest load current will be drawn by the load. But at the highest operating frequency, the highest average RMS ripple currents will have to enter and leave the filter capacitors. So one should calculate the required capacitor value at the highest and lowest loads and use the higher value [see Eq. (12.5)].

$$C_{out}\ (+5) = 100\ \mu\text{F}$$
$$C_{out}\ (+/-12) = 10\ \mu\text{F}$$

Ideally these capacitors should be very low ESR/ESL glass or ceramic capacitors. Unfortunately, to attain these values a large number of them would have to be paralleled. Plus they are very expensive. Instead, one can use a few expensive low-ESR/ESL capacitors and place them between the filter choke and some less expensive tantalum capacitors. The high-frequency capacitors will filter the high-frequency current components, and the less expensive tantalum capacitors will provide the bulk storage for the load.

12.3.6 The Output Filter Chokes

The output filter chokes are determined in the same way as done in the previous example, except the maximum "off" period has changed. If one substitutes $1/f(\text{min})$ for $T_{\text{off}}(\text{max})$ one obtains:

$$L_{\text{out}}(\text{min}) = \frac{V_{\text{sec}}(\text{RMS})}{1.4(f_{\text{min}})(I_{\text{out}}(\text{min}))} \qquad (12.31)$$

$$L_{\text{out}}(+5) = 43.6 \ \mu\text{H}$$

$$L_{\text{out}}(+/-12) = 410 \ \mu\text{H}$$

12.3.7 Driving the Power MOSFETs

Now because the power MOSFETs are being driven at a much higher frequency, the power dissipated within the output drivers of the control IC become prohibitive, especially at elevated ambient temperatures. It has become necessary to add external drivers to the control IC. This can easily be done by selecting nearly complementary *NPN* and *PNP* small signal transistors and wiring them as a common-emitter totem-pole driver. The emitters follow the voltage on the base, but the current is amplified by their current gains. Power MOSFET driver ICs such as the MC34152 can also be used for this purpose.

12.3.8 Control Circuitry Considerations

1. The use of a quasi-resonant controller IC makes it easy to design the controller section of the power supply. This example will use the Motorola MC34066P.
2. From the graphs and equations contained within the data sheet, the values for the timing elements are easily determined (so I won't bore you with details here).
3. *Note:* Special attention should be given to the separation of "digital" (or switching) grounds and analog grounds (amplifier-related grounds). One point grounding practices dictate that analog grounds should center around the base of the current-sensing resistor. Then all primary grounds should be tied to the source lead of the lower MOSFET.
4. The compensation of the error amplifier should now be recalculated using an amplifier bandwith of 40 kHz (or 200 kHz/5). This will

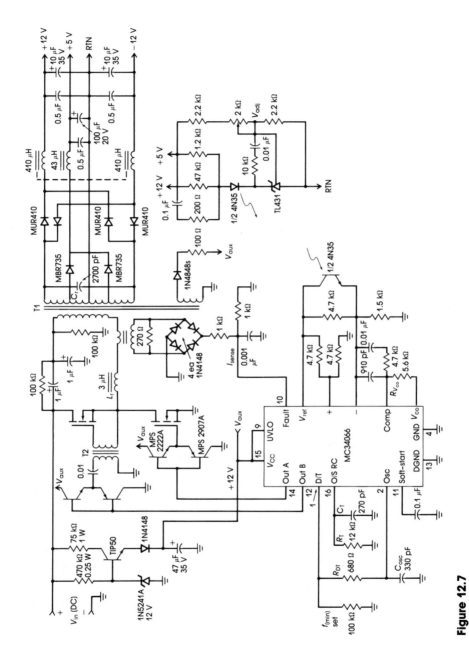

Figure 12.7

A 50-W, 1-MHz, zero-current-switching, quasi-resonant, half-bridge converter.

yield a much better transient response. The procedure for determining the new values of the two-pole–two-zero compensation method are identical to the PWM half-bridge example. (*Note:* For some operational amplifiers available gain at 40 kHz may be insufficient to provide the desired loop bandwidth. In this case it may be desirable to add an external, high-gain-bandwidth operational amplifier such as the MC33281P and use the controller amplifier as a noninverting buffer.)

12.3.9 Construction Recommendations

1. Use wide PCB traces for the high current paths within the board. This will reduce the series inductance and resistance of the trace. It will also reduce the RF radiation.
2. Keep high-current paths as short as possible!

12.4 A 60-W, Off-Line Flyback Converter with Battery Backup

This converter is actually in use in office PBX telephone systems. It demonstrates the flexibility within the designs of the transformer isolated switching power supplies. The illustrated design procedure has been abbreviated because it would be redundant with the other design examples.

12.4.1 Design Specification

Input Voltages

AC: 90 to 130 V AC   USA, Canada, and Japan

          200 to 240 V AC   Europe

DC: -40 to -60 V DC

Output Voltages

$+5$ V DC at 1.25 A   (0.25 A minimum load—variable)

-5 V DC at 0.25 A   (load is constant)

-48 V DC at 1.00 A   (load is constant)

Line and Load Regulation

$+/-5$ V DC: 0.1 V variation maximum

-48 V DC: 2.0 V variation maximum

Safety Considerations

The AC sections of the supply will be isolated from the DC (outputs), and -48-V sections of the supply and will meet VDE specifications.

Backup Considerations

The supply will be able to start from either the AC or DC input. When operating from the AC input, there will be no current drawn from the DC input. The supply will switch to the DC input automatically on the absence of the AC input. The DC input may not necessarily be present at all moments in time.

12.4.2 Predesign Considerations

1. Since this unit is going to be used internationally, VDE imposes the most stringent safety requirements. These requirements will be the ones used in the design of the power supply.
2. The AC input circuitry and any auxiliary power windings will have to be isolated to 3750 V AC, including creepage and clearance.
3. The battery backup states of operation are shown below.

AC	DC	AC Circuit	DC Circuit
On	On	Operating	Inhibited
Off	Off	—Nonpowered state—	
On	Off	Operating	Inhibited
Off	On	Inhibited	Operating

   So some form of optically isolated logic will have to be incorporated into the startup and auxiliary power circuits to produce these operating states. Accuracy is required for the sensing of the loss of AC so that a clean, uninterrupted switchover can be done.
4. The $+5$-V output is highly variable, so cross-output sensing should be employed to improve the cross-regulation of the supply.
5. With the isolation requirements, completely separated controllers and startup circuits will have to be used. If isolation of both the primary circuits were not required, the functions could have been merged.

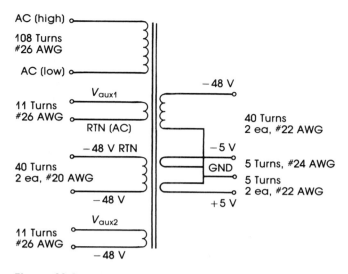

Figure 12.8

Transformer design for a 60-W off-line flyback converter with battery backup.

6. Since cost is a critical issue, it is desirable to use one supply instead of two. This means that both the AC and the DC will have separate controllers and primary windings on one transformer.

7. The power requirements allow a flyback converter topology to be used that will yield the smallest parts count and cost (see Fig. 12.8).

8. A voltage doubler AC line rectifier and filtering circuit was used since the supply board was designed for international use.

12.4.3 Transformer Design

Refer to Section 6.3 for the design procedure for a flyback transformer.

1. A pot core is to be used because of its shielding properties. The one selected was an F42616 with a 33-mil gap.

2. A frequency of operation of 25 kHz was selected mainly because it is desired that the $+/-5$-V-DC output needs a minimum of one turn. Those particular windings drove the entire design of the transformer. The transformer was designed from the back to the front. Several iterations were required to set the turns for each winding and the frequency of operation.

3. $+12$-V auxiliary power windings were required for each controller

to enhance the efficiency of the supply. The auxiliary windings are isolated along with the primaries.

The resulting transformer was as shown in Figure 12.8.

12.4.4 Other Design Considerations

1. Two almost identical current-mode controller circuits were used. The differences were in the startup circuits and the current-sensing resistor values.
2. The voltage feedback consisted of two optoisolators where the LED diodes were placed in series to sense the same current, and the output transistors went to their respective controllers.
3. The negative outputs were sensed, since it was desired to use the cross-sensing technique outlined in Chapter 7. Because of variations in the optoisolator current transfer ratios, it was necessary to adjust the regulated outputs at each controller. This was done via a selected resistor at each controller's error amplifier at the time of final production test.
4. Diodes will be added to the drains of each MOSFET in order to isolate their respective winding when that controller is in the inhibited mode. This prevents the intrinsic diodes from conducting during the other MOSFET's conduction cycle and connecting the winding to the discharged input filter capacitor.

Figure 12.9
AC rectifier/filter section.

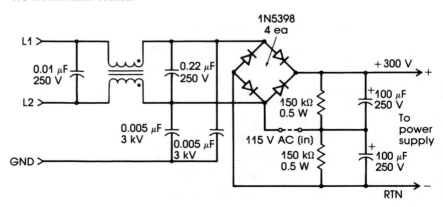

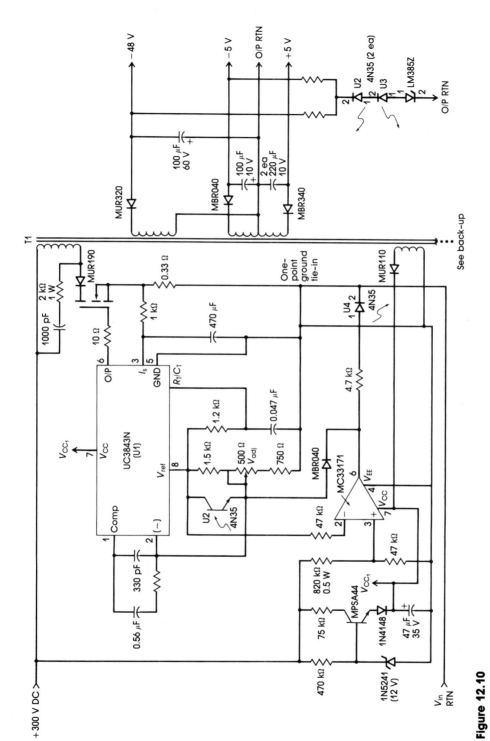

Figure 12.10

High-voltage controller with outputs.

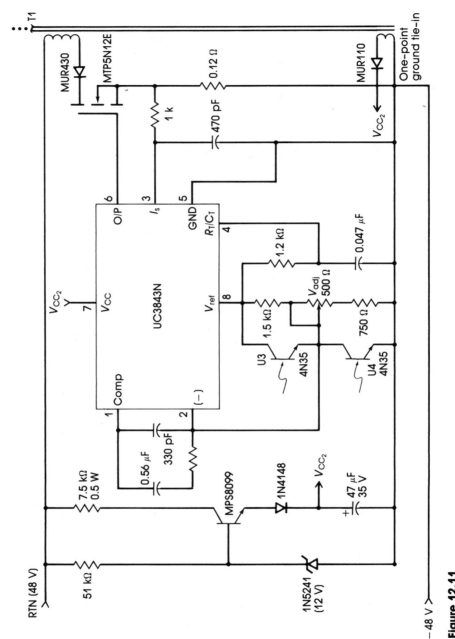

Figure 12.11

Low-voltage controller (battery backup).

12.4.5 Postpaper Design Changes

1. The transformer core was changed from the pot core to an E core for two reasons; (1) The windings with insulation did not fit easily into the window area, and (2) the cost of the finished transformer was too high. The E core is the least expensive core and is inexpensive to wind. It also had a larger winding area. To do this, I found an E core size with approximately the same cross-sectional core area and magnetic length. The off-the-shelf gap was different, so the turns had to be recalculated. There was no problem in fitting the windings within the window area. Plus a Faraday shield had to be added to the exterior surface of the windings. The transformer was substituted without a problem.
2. The resultant cost was $27 each in 100 thousands quantities.
3. The final effiency was:
   50 percent of rated load: 81%
   100 percent of rated load: 78%

See also Figures 12.9, 12.10, and 12.11.

Bibliography

Alberkrack, Jade (1984). "Theory and Applications of the MC34063 and uA78S40 Switching Regulator Control Circuits" (AN920A). Motorola, Inc.

Brown, Martin J. (1988). "Switching Power Supply Design," seminar notes. Motorola Semiconductors Products Sector.

Carsten, Bruce (1986a). High frequency losses in switchmode magnetics, High Frequency Power Conversion Conference, May.

Carsten, Bruce (1986b). Switchmode design techniques above 500 KHz, High Frequency Power Conversion Conference, May.

Chryssis, George (1986). "High Frequency Switching Power Supplies," pp. 88–104; 181–194. McGraw-Hill, New York.

Dash, Glen (1987). Designing for EMI compliance—Part 1 designing the PC board, *Compliance Engineering*, pp. 71–80.

Dash, Glen (1987). Product liability—Technology and the insurance crisis, *Compliance Engineering*, pp. 164–173.

Dash, Glen (1987). Selling safely overseas, *Compliance Engineering*, pp. 178–184.

Dash, Glen (1987). Focus on power supplies, *Compliance Engineering*, pp. 185–186.

Dixon, Lloyd H. (1986). Closing the feedback loop, "Unitrode Power Supply Design Seminar," pp. cl-1–cl-31.

Estrov, Alex (1986). Power transformer design for 1 MHz resonant converters, High Frequency Power Conversion Conference, May.

Gauen, Kim (1983). "Designing with TMOS Power MOSFETs" (AN913). Motorola, Inc.

Jachowski, Mike (1988). The use of new high speed amplifiers in the control of switchmode power supplies, High Frequency Power Conversion Conference, May.

Javanovic, M. M. (1987). Design aspects of high frequency off-line quasi-resonant converters, Virginia Polytechnic Engineering Conference, September.

Kepple, Niel (1977). High power flyback switching regulators, WESCON, September 20.

Kirchdorfer, Josef (1987). Overcurrent protection—The hurdle of differing standards, *Compliance Engineering*, pp. 189–190.

Lee, F. C. (1987a). Zero-voltage switching techniques in DC/DC converters, High Frequency Power Conversion Conference, April.

Lee, F. C. (1987b). High frequency quasi-resonant converter topologies, Virginia Polytechnic Engineering Conference, September.

Lee, F. C., and Liu, Kwang-Hwa (1986). High frequency quasi-resonant power converters, High Frequency Power Conversion Conference, May.

Liu, Kwang-Hwa, and Lee, F. C. (1987). Zero-voltage switching techniques in DC/DC converters, Virginia Polytechnic Engineering Conference, September.

Magnetics, Inc. Design manual featuring tape wound cores.

Magnetics, Inc. Ferrite cores—catalog.

Magnetics, Inc. Powder cores—catalog.

Ridley, R. B. (1988). Multi-loop control for quasi-resonant converters, Virginia Polytechnic Engineering Conference, September.

Ridley, R. B., Tabisz, W. A., and Lee, F. C. (1988). Multi-loop control for high frequency quasi-resonant converters, High Frequency Power Conversion Conference, May.

Schultz, Warren (1982). "Power Transistor Safe Operating Area—Special Considerations for Switching Power Supplies." Motorola Semiconductor Products Sector.

Venable, H. Dean (1983). The K factor: A new mathematical tool for stability analysis and synthesis, POWER-CON10, March.

Index